A SHEARWATER BOOK

HEARTSBLOOD

But it's poets who cry
on warm summer nights
remembering the cold to come.

—"beringer"

HEARTSBLOOD

HUNTING, SPIRITUALITY, AND WILDNESS IN AMERICA

David Petersen

Island Press / SHEARWATER BOOKS
Washington, D.C. • *Covelo, California*

A Shearwater Book
published by Island Press

Copyright © 2000 David Petersen

All rights reserved under International and Pan-American Copyright Conventions.
No part of this book may be reproduced in any form or by any means without permission
in writing from the publisher: Island Press, 1718 Connecticut Avenue, N.W., Suite 300,
Washington, DC 20009.

Shearwater Books is a trademark of The Center for Resource Economics.

Library of Congress Cataloging-in-Publication Data
Petersen, David, 1946–
    Heartsblood: hunting, spirituality, and wildness in America / David
Petersen ; foreword by Ted Williams.
        p.   cm.
Includes bibliographical references and index.
    ISBN 1–55963–761–7 (cloth : alk. paper) — ISBN 1–55963–762–5 (paper :
alk. paper)
    1. Hunting—Moral and ethical aspects. 2. Hunting—Philosophy. I.
Title: Heartsblood. II. Title.
SK14.3 .P48 2000
179'.3—dc21   00–009400
                                                                   CIP

Printed on recycled, acid-free paper ✪
Manufactured in the United States of America
10 9 8 7 6 5 4 3 2 1

*For Paul Shepard
and Florence Rose Shepard*

Also by David Petersen

Elkheart: A Personal Tribute to Wapiti and Their World

*The Nearby Faraway: A Personal Journey
Through the Heart of the West*

A Hunter's Heart: Honest Essays on Blood Sport (editor)

Ghost Grizzlies: Does the Great Bear Still Haunt Colorado?

Earth Apples: The Poetry of Edward Abbey (editor)

*Confessions of a Barbarian: Selections from the
Journals of Edward Abbey* (editor)

*Racks: The Natural History of Antlers and the Animals
That Wear Them*

Among the Aspen

*Big Sky, Fair Land: The Environmental Essays
of A. B. Guthrie, Jr.* (editor)

Among the Elk

Contents

Foreword

T HERE WAS THINGS which he stretched," as Huck said of Mr. Mark Twain, "but mainly he told the truth." People never read forewords first, so you may already know about me a little from Mr. David Petersen. One of Mr. David Petersen's stretches was that I should have written this book. Lucky I didn't.

I get annoyed when thinking about how fish and wildlife are managed, pursued, harvested, and especially, "defended" by people who know virtually nothing about them, people who have never been close to them in their natural habitat, people who have never bothered to learn their true nature or appreciate and love them for what they are. Anger impedes the framing of convincing arguments, so I would have given you a screed—perhaps a stylish, colorful one, but at best you would have been entertained, not educated. And in the end you would have been no better prepared to frame your own convincing arguments, of which I have heard none until this fine book.

After working for a state game and fish agency and then as managing editor of *Gray's Sporting Journal*, I had despaired of ever reading anything original, helpful or even interesting on the subject of why hunting is, or can be, socially respectable, and therefore I had vowed never again to try. Fortunately I had to break that vow in order to complete this assignment.

I have never known a more passionate hunter than David Petersen, so I'm confused, impressed, and envious about his ability to write dispassionately on this subject. Not that there isn't plenty of passion in his book, but when Petersen needs to turn it off and step back and analyze, he does so. He writes like he hunts—without baggage.

Hunting and writing are pretty much the same thing, anyway. You do both by watching, listening, learning, exercising patience, acquiring humility, and always trying to get it exactly right. You pursue the elusive word as you would the elusive elk.

When education of sportsmen was my job I wondered why neither I nor my colleagues could ever succeed at it. Now I know: we should not be educating sportsmen because if we need to, it is usually already too late. We instead should be focusing our energies on educating children. Because if a hunter, to use Petersen's example, stops his car on a road and lobs an arrow at a game animal 100 yards away, there's probably no hope for him. You can tell him that he could get scolded or, depending on the state, cited for such behavior, and he may desist in the future, but he won't understand what hunting can be unless by some miracle he has a self-induced epiphany. As Petersen writes, "Rather than creating personalities and worldviews, hunting merely reflects them, good and bad, as shaped by the overarching human environment."

America and the world need to take a few pages from Petersen's book—that is, learn about nature in the wild instead of on the Discovery Channel. Last May I sat with several dozen people at an outdoor party as twilight settled over a northern hayfield. Wood thrushes were caroling from chartreuse hardwood and hundreds of fireflies were flashing their green lanterns over the high grass. For anyone who loves wild things and wild places, the scene was worthy of note. But other than my wife and me, everyone seemed to be utterly oblivious to what was happening all around us. Most of them probably do love wildness (or at least the idea of wildness), but it was almost as if they needed a tour guide to tell them it was time to "do nature." As a wise wildlife educator once told me, "You can't 'teach' anyone anything. You can only provide opportunities for learning." The opportunities are all around us, but most of us don't take them. We look for nature everywhere but where it exists.

Maybe Petersen's ability to reason calmly and communicate clearly on the subject of hunting, animal rights, and wildlife management comes of spending so much time outdoors watching wild animals and trying to figure out what it is they're doing and why. As his wife says, he's beyond the point of just wanting to kill or even watch elk; he wants to *be* one. Petersen is a fine writer, despite his protestations to the contrary, because he writes a lot, he works hard at it, and mostly, because he's a keen observer. Petersen's graceful and witty prose makes this book a fun read, but it's his coolness and clarity of thought that make it a *useful* read.

Heartsblood may not change your mind about hunting, but that's not its purpose. What it can do is arm you with compelling arguments (should you want them), show you who the real enemies of fish and wildlife are and are not, and help you learn about nature as it really is instead of the way some-one, anyone—even Petersen—says it is. This book can also help you provide learning opportunities for others. Think about Petersen's message. Then put this book down and learn on your own—hunting, fishing, counting coup, or just watching. Get out there with fish and wildlife. *Do nature.* The only equipment you'll need will be open eyes, open ears, and an open mind.

Ted Williams,
Editor-at-Large, *Audubon* magazine and
Conservation Editor, *Fly Rod & Reel*

Preface
Contemplating *The Hunter*

T HERE'S SOMETHING HAUNTING, pensive, and, to most North Americans today, incomprehensible about that feral, near-naked fellow gracing our cover. His name is *The Hunter* and he was created—conceived and painted to life—by N.C. Wyeth back in 1906, not so long after such scenes were as common as freeway commuting today. A framed print of this golden-age, golden-morn, iconic Noble Savage hangs in my little writing shack (a.k.a. the Outhouse) here in the Colorado Rockies, and I reflect upon it often.

What is it, exactly, that Wyeth means to imply via *The Hunter's* enigmatic expression? With the proverbial bird in hand—a Canada goose, having been "reduced to possession"—what's the meaning of the longing look *The Hunter* is shooting, not unlike an arrow, at the high-flying flock V-ing across that ethereal crepuscular sky?

Might *The Hunter's* expression suggest a visceral, tooth-and-claw hunger for meat, tied in a primitive savage knot with an innate (male) lust for death, destruction, and domination—as the most ardent of contemporary anti-hunters choose to believe of all contemporary hunters?

Or might *The Hunter's* gaze reflect a hunger of a different sort entirely?

———

T O THE ATTENTIVE EYE and searching mind—the hunter's eye and mind—there's more said in and by Wyeth's painting, and said better, about freedom and wildness and life and death and deep-rooted universal order and *truth,* than all the words in this or any other book, no matter how informed, philosophical, or "holy," can ever hope to convey. But no matter. Inspired by *The Hunter* and others, as well as my own life experiences, I intend to try.

Try what? Try to establish where and how we got to where and how we now are as a species.

Try to draw clear distinctions between *hunting* and *hunter behavior*—praising what's right about the former and damning what's wrong with the latter.

Try to render the labels "hunter" and "antihunter" palpable—to put faces on these much-used but little-understood generalizations.

Try to identify that slippery thing we call "spirituality."

Withal: Try to *get it right*.

Which is to say: The meandering thoughts and unabashed opinions that fill the following pages aren't just about hunting and "animal rights," not by a long shot. What they're about is life's basic truths, as best I can ferret them out and whether I, or you, always like the results or not. And almost as often as not, when it comes to hunting, I don't like it one bit. But that changes nothing—which is precisely the point of all to come.

And now for some well-earned. . . .

Acknowledgments

THE INFLUENCE that the "father of human ecology," Dr. Paul H. Shepard, has had on my thinking and living, and thus on the ruminations that fill these pages, is evident: I openly rely upon his wisdom and words throughout. Thank you, professor. How I wish we could have shared some campfires.

Happily, I have shared campfires with Florence Rose Shepard (a.k.a. "Philosophy Rose"), Paul's wife and editor, a brilliant thinker and writer in her own right. Flo, by any name, you are a Rose indeed.

Another intellectual and lifestyle mentor—and one I shall continue to acknowledge in every book I ever write—is my old friend and exemplar, Edward Abbey. Ed: we have not forgot.

Yale professor Stephen R. Kellert has been most generous in allowing me to quote liberally from his meticulous studies of human perceptions of wildlife—and more generous yet in his constructive criticisms of my interpretations of the broader and deeper implications of those studies.

Only after I'd completed the initial draft of this book did I discover *Wildlife and the American Mind,* by Mark Damian Duda et al., of Responsive Management, a Virginia-based public opinion polling and research firm specializing in wildlife, natural resources, outdoor recreation, and environmental issues. In reading through those eight hundred informative pages, it was gratifying to discover how often Responsive Management's scientific research has produced statistics that agree with my own empirical (that is, decidedly nonscientific) experiences and personal conclusions. I am grateful for RM's generosity in allowing me to sprinkle the statistical results of their good work throughout this text.

For their generosity in providing expert information and candid criticism throughout my researching, writing, and revising, I warmly thank my friends, advisers, and betters: Tom Beck, Mike Buss, Valerius Geist, Dave Stalling, Ted Williams (who could and probably should have written this book, rather than just the foreword), and many unnamed but not forgotten others. The facts relayed in these pages are hard and true and largely theirs; all interpretations and opinions, unless otherwise noted, are my own.

Likewise, my handlers at Island Press—the inimitable and indefatigable Barbara Dean, the exacting Barbara Youngblood, the didactic Don Yoder, and the affable Christine McGowan—all are good as gold. (Better in fact, as none of us is in this for the shekels.)

It's with boundless gratitude (Shakespeare's "exchequer of the poor") that I thank Caroline, to whom I owe the twenty best years of my life. My thanks also to Clarke Abbey, John Corcoran III (hold on, brother; we're coming), Dan Crockett, Thomas Aquinas Daly, Lane Eskew, Erica Fresquez, Carl Brandt, Pamela Johnson, Scott Stouder, Mitch Caldwell, Ken Wright, Bob Yetter—and of course *The Hunter,* that fiercely feral, modestly clad wildman on the cover, who appears here by permission of the Brandywine River Museum, Chadds Ford, Pennsylvania (elegant repository of N. C. Wyeth's art and spirit).

Scattered bits of this book saw first publication in the indulgent pages of *Backpacker, Bugle, Inside/Outside, High Country News, MountainFreaks, Mule Deer, Northern Lights, Sports Afield,* and *Wild Duck Review* and in my foreword to Paul Shepard's 1999 essay collection, *Encounters with Nature.* Other issues and ideas explored in the following pages found their inspiration in verbal harangues I delivered to Ontario's Hunting Heritage/Hunting Futures conference (Toronto, August 1998) and at the Wilderness and Spirituality lecture series sponsored by the Wilderness Institute of the University of Montana (Missoula, April 1999).

Muchas gracias, compadres. May the spirit run with you all, forever.

D. P.
from the Outhouse
San Juan Mountains, Colorado

PART I

Exploring Hunting's Heritage

For many, more and more of us, the out-of-doors is our true ancestral estate. For a mere five thousand years we have grubbed in the soil and laid brick upon brick to build the cities; but for a million years before that we lived the leisurely, free, and adventurous life of hunters and gatherers, warriors and tamers of horses. How can we pluck that deep root of feeling from the racial consciousness? Impossible.

—Edward Abbey

1

How Hunting Made
(and Helps to Keep) Us Human:
A Tribute to Paul Shepard

> When I was a caveman
> painting on the walls
> I never had a dollar
> but man I had it all.
>
> —Chris Smither

ONE OF MY MOST CHERISHED TEACHERS, the late Paul Shepard, liked to point out that "in defiance of mass culture, tribalism constantly resurfaces." The minority of hunters who are true (to their heritage) and authentic (in their goals and actions) form just such a tribe, though many hunt alone.

———

IT'S MID-AUGUST and hot here in southern Colorado. Just two more weeks until the opening of archery elk season, which I await each year with all the patience of a child counting down the days to Christmas. To help me through this anxious time, to facilitate getting in shape for the vertical challenge of elk hunting and just because I like it, I'm spending a few days alone, out among the crafty creatures locals know as "prairie goats," albeit with no real hope of bringing home anything meatier than memories. As one frank friend once observed, stalking pronghorn with two sticks and a length of string is about as productive as chasing farts in a hurricane.

So true. And so what? In three years of hard trying, I've yet to work close enough to a wily pronghorn to loose a conscionable arrow. No matter. In hunting as in life, it's essential to avoid tunnel vision. Plenty good enough just to be alive, free, healthy (so far as I know), and semi-sane in a world gone largely mad.

To the point: It's a multiple blessing just to be here now, in this quietly spectacular high-desert place, so near to home in miles, yet so geologically and thus spiritually foreign: a Nearby Faraway if you will (with apologies to Georgia O'Keeffe). How different this place is from the lofty, lime-green mountains I call home, over on the far side of the Continental Divide—visible even now, off in the western distance, quilted here and there with raggedy hold-out patches of last winter's snow, shining cool as a promise beneath a roaring late-summer sun.

Well, that promise—of dark-forested mountains animated by bugling bull elk in their piss-perfumed rutting wallows beneath brassy autumn aspens rattling in a cool mountain breeze—all of that and more will just have to wait. For now, this rocky pronghorn heaven will do just fine.

Not your typically pan-flat, cow-burnt, gnat-tortured prongy stamping ground, this, but scenic basin-and-range country: rolling, rugged, vulcanized and cliffy, punctuated with verdant pockets grown waist-deep in aromatic sage and yellow-flowering rabbitbrush. Average elevation: eight thousand four hundred feet above the far-off sea.

This is national forest land, mostly, just like at home. But here, ironically, there is no forest; none of the "timber-quality" trees the USDA Forest Service so loves to cut and sell well below cost, at taxpayer expense, ripping the landscape with (taxpayer-subsidized) roads while they're about it.

In light of the lack of logging, the federal keepers of this sparse land manage to find other uneconomical, unecological "uses" for it, allowing it to be sheep-burnt just this side of hell every spring by a brainless blight of woolly maggots.

Domestic sheep, here and everywhere they roam throughout the semi-arid West, overgraze the vegetation, foul the water, denude riparian areas, and import noxious invader plants—thistle, leafy spurge, and the like—that squeeze out native vegetation, thus poisoning and starving the ecosystem. Additionally, domestic sheep harass and even kill wildlife (coyotes, bears, lions, and wolves in particular) with help from gun-toting ranchers and taxpayer-subsidized government trappers, spread disease to wild animals (especially bighorn sheep, of which a secretive few lurk hereabouts), and otherwise displace wildlife.

And more. All of it bad for nature. For wild*ness*. Bad for you and your children, and devastating to the feral likes of me. Private livestock belongs on private range, not on America's nominally communal, nominally "wild," public lands.

But let's forget the stinking sheep for the moment. What else defines this place? Prickly pear here and there, mostly in whiskered clusters; claret cups and spearlike yucca; too many ugly dirt roads with their cruising patrols of pathetic road warriors—pretend hunters looking for easy targets to plug, illegally, from their vehicles. (A losers' game, start to finish, but a popular game nonetheless.) Knotty little pinon pines are ubiquitous all across this rocky, semidesert landscape, though curiously you won't see a one of their sympatric, symbiotic sylvan sisters, genus *Juniperus,* a shag-barked juniper known locally, and incorrectly, as "cedar."

Even so, despite the scars of the long-term, private-interest, coldly commercial abuses the USDA Forest Service euphemizes as "multiple use," this is one of the loveliest places I know. And much of its beauty arises from its eerie ambiance. The view all around is long and lean, framed on three sides by a majesty of rocky mountains—San Juans south and west, Sangre de Cristos (sweet Blood of Christ!) eastward—and in the near north by stark volcanic cliffs the color of cowboy coffee.

Additionally, owing to this anomalous high-desert basin's altitude and proximity to major mountains, brief violent thunderstorms erupting from massive dark clouds—great steely-blue anvils, pregnant with pyrotechnic energy—come grumbling and flashing through most every afternoon in summer, blessing this parched place with a benediction of chiaroscuro (visible-beam, some call it "God") light.

To many who know and appreciate it, this arid, parsimonious landscape evokes the Alaskan tundra. To me, the travel this place implies is more in time than in miles, all the way back to the icy old Pleistocene, that finishing ground of human evolution.

———

BACK THEN, twelve thousand years ago, with North America just emerging from the long icy ages, the people living here were seminomadic spearchuckers with such provincial names (assigned by us) as Clovis, Folsom, and

Cody. Perhaps, as long believed, they were the recent descendants of wandering Mongol hunters who emigrated overland, via the Bering land bridge, and quickly spread throughout North America. Or perhaps, as science and I are coming increasingly to believe, they were "caucasoids" of the Solutrean persuasion who immigrated here much earlier, more than eighteen thousand years ago, by land and sea, from Europe's Iberian Peninsula (comprising Spain, Portugal, and southwestern France). And just as likely as either of these scenarios, they were some of both. Right now, no one knows with absolute certainty. Nor does it matter all that much to this discussion.

What does matter is that these early Americans were full-time hunter/gatherers who earned good honest livings by killing, dismembering, and hungrily devouring mammoths and other Pleistocene megafauna.

Too soon, however, these gorgeous gargantuan prey species met their inevitable end—due in largest part, likely, to radical climate and (thus) habitat changes, possibly exacerbated by microbial diseases imported by in-migrating humans . . . and yes, in the end, by hunting pressures exerted by those same two-legged invaders.

With the mammoths gone, the agile subsistence hunters turned their predatory attention to giant wide-horned bison, which came to drink at glacier-gouged potholes filled with lucent meltwater. The stones forming these ancient ancestors' fire rings and the coals of the fires those rings contained, along with flaked stone butchering tools and the tool-scarred bones of prey species, lie scattered hereabouts in shallow-buried abundance.

Aside from being drier, the looks of this place haven't changed much. Consequently—owing to the deep-time, subtly magical ambience of the landscape—my hunts, camps, and hikes here have evolved into an annual pilgrimage.

It follows naturally that every time I come here, every year at just this time, my thoughts run to one of the primary influences in my spiritual, intellectual, and hunting life. I'm referring to Dr. Paul Shepard: Pleistocene prophet, prehistoric philosopher, postmodern primitive, and father of the science of human ecology, the study of humanity's timeless, coevolutionary, developmental relationship, both as a species and as individuals, alongside our fellow wild creatures in shared wilderness settings.

In Shepard's vision—in fact a radical revision of the standard take on our "savage" human beginnings—the human hunting heritage leaps to lively and meaningful life. Attempting to simplify and summarize the work of a scholar so compelling, controversial, and complex as Paul Shepard, which even many academics find challenging—and, further, to suggest how such thinking applies to modern "sport" or "recreational" hunting—well, it doesn't make for casual campfire conversation. Yet the exercise—the hard intellectual work necessary to comprehend Paul Shepard—is worth the effort. That's because he, more than any other thinker I've encountered in half a century of searching, offers answers to the central query of this exploration: Why do hunters hunt?

Nor is this so esoteric a question as it may at first sound. As Shepard reveals, in the bigger picture we might as well ask: Why are we humans, and not still apes?

———

TRUE HUNTERS, whatever their differences in geography, culture, gender, or experience, are kindred spirits, a tribe united by a shared love for what Paul Shepard, employing double entendre, calls the "sacred game"—in my present instance, the *pronghorn* is the sacred game; the pronghorn *hunt* is the sacred game. It's a love, this ancient game, that cuts so deep yet remains so inexplicable as to seem almost instinctive.

Why so? What is the ultimate source, the prime mover, of the human animal's clearly intrinsic *yearning* to hunt and fish? Why do young children, across all cultures, respond more warmly to animals than to any other class of objects, real, toy, or "virtual"? (The current Pokémon rage is commercial culture's strikingly successful attempt to capitalize on this childhood interest in real animals by morphing them into toys, video games, collector cards, and TV cartoons. The spiritual loss to the children thus victimized is, for the moment, incalculable.) Why are the names of animals and the sounds animals make among the first words uttered by most human infants—again across all cultures? ("Because," you may well reply, "there are so many kid's books, toys, games, and TV shows based on animals today." But, to press the point, why is this so?)

Likewise, why is the color forest-green universally perceived as restful and reassuring, while blood-red universally excites and agitates?

Finally, circling these universal biophilic feelings back to where we began: What hidden engine drives the more superficial motives, those things so often named as reasons for modern hunting: meat, adventure, companionship, challenge, antlers, and other so-called "trophies"?

What indeed? And why? Why do so many modern humans still hunt, when so few of us really *need* to hunt in order to physically survive?

———

MOST MODERN HUNTERS, good and bad, just want to hunt—not explore and debate why they do it and how they do it and what others think of them for it. Yet today, no thoughtful hunter can afford to just hunt. In order to defend what we do—to ourselves, our families, our friends, and, especially, to an increasingly urbanized, denatured, domesticated, and virtualized populace—in order to improve hunting ethics and invite and inspire tomorrow's hunters and assure that hunting *has* a tomorrow . . . for all of these reasons and more, hunters *must* ask themselves why.

And we must answer honestly. For only by discovering the whys can we hope to influence the hows. And only by reforming the hows, where necessary, can we effect the essential change without which modern hunting is likely doomed, and likely deserves to be.

———

I GREW UP IN OKLAHOMA. It was a place—back in the "olden days" of my midcentury youth at least—where hunting and fishing were readily available to anyone who wanted them. Rabbits and squirrels and quail and catfish and bass all lurked just a bicycle ride from my city-edge home. ("Suburban" was not quite yet a viable term.)

Likewise, in that place and time, so-called blood sport was roundly accepted. While not everyone hunted, almost everyone had family or friends who did, while others kept chickens or rabbits or maintained close contact with relatives on the farm. Simple Okies we were. Yet we possessed a basic sort of knowledge largely lost today: We knew where our food came from.

I recall wondering, even back then, even as a child, why I felt such an all-consuming *need* to hunt. It has never been something I've taken for granted. It wasn't just for meat, though almost from the start I ate everything I killed. It wasn't just for fun, though fun it was: the camping, the hiking, the learning, the adventure. It wasn't and never would be for "trophies," at least not in the macho/ego sense of that multifaceted and massively misunderstood term. Certainly, companionship played a part: the best of my friends all were hunters.

And yet, from the day I could legally drive, I've hunted mostly alone. Nor did I have more than hit-and-miss mentors to nourish my taste for wild meat, wild country, and the wild life of the chase. For me, from the beginning, hunting has been not a family thing, not a peer thing, not a cultural thing, but a self-powered passion—like a bird's urge to fly, a fish's fervor to swim . . . like *instinct.*

Naturally curious about why I do what I do, for decades I've chewed this quandary: Could there be such a thing as a human hunting instinct, a genetic vestige from our long, formative hunting/gathering past? And if there is, why does this incipient instinct flame so hot in some while seeming dead as cold ashes in most and even infuriating a few?

Although it was fun to think about this "hunting as instinct" hunch, definitive answers remained always just beyond my intellectual grasp. Then, finally, I discovered Paul Shepard—who explained my life to me. Since Shepard plays such an important role in the pages to come—I quote, paraphrase, and otherwise rely on him more than any other single source (save perhaps my personal experience)—allow me to profile the man and outline his philosophical science of human ecology, best as I can, right here and now.

———

IN PARALLEL CAREERS as scientist, scholar, teacher, and writer, Paul Shepard spent an actively joyful life examining the skin-tight fit between human nature and wild nature, as revealed in the titles and subtitles of his books and the personal philosophical progression these titles reflect: *Man in the Landscape, The Tender Carnivore and the Sacred Game, Nature and Madness, The Others: How Animals Made Us Human, Coming Home to the Pleistocene, Encounters with Nature,* and more. Driving all of this was Shepard's exhaustive inquiry into how millions of years of evolution under the tutelage of full-time hunting, scavenging, and gathering—together termed "foraging"—made us human.

Paul Shepard was born in 1925 and grew up in the Missouri Ozarks. During World War II he landed with the Twelfth Armored Division at Normandy and took part, as an artilleryman, in the Battles of the Little Bulge. After the war, Shepard took a doctoral degree in ecology and art history at

Yale University and enjoyed a long and distinguished career as a research professor of human ecology—a new way of investigating the long-accepted truth that "all things are connected."

Somewhat ironically, considering his keen intellectual interest in hunting, Shepard was not himself devoted to the activity. Rather, like José Ortega y Gasset ("the hunting philosopher"), Henry Thoreau, naturalist Ernest Thompson Seton, and so many others, Shepard *had* hunted—and considered the experience important, if not essential, to obtaining a balanced, biological worldview.

As a Missouri country boy, in company with friends, young Paul hunted small game and birds and ran a trapline for rabbit and muskrat. As a young adult, he trained and flew hunting hawks and falcons. And—a classic hunting yarn by any standards—while serving with the Army of Occupation at Heidenheim, he ducked out one day to hunt deer, only to flush a hidden covey of German soldiers, whom he escorted back to his unit as prisoners.

Though not well known beyond academic and scientific circles—he never sought pop celebrity and shunned fame when it sought him—Paul Shepard is widely hailed as among the finest minds of the century, of *any* century; a revolutionary freethinker who envisioned a whole new way of looking at the origins and meaning of life on earth and then conducted a lifelong hunt for unimpeachable scientific, philosophical, and empirical evidence to back that vision.

The radical new take on human history and evolution envisioned by human ecology has been called "the subversive science," by supporters as well as critics, because it shakes the essential underpinnings of civilized culture. Its primary focus is the evolutionary interplay between wild animals, wild landscapes, and preagricultural (foraging) people, and how those ancient mutual dependencies worked their way into the human DNA wiring diagram, or genome, influencing us all still today.

Gathering scattered nuts of knowledge from a broad field of scientific research—archaeology, anthropology, ethnology, ethology, biology, psychology, sociology, evolution, genetics, and more—Shepard synthesized it all to show how millions of years of vocational hunting and gathering, along with the small-group, nomadic, clan, and tribal lifestyle demanded by such a lifeway, made us what we are today—for better and (when our genetically prescribed needs are neglected or perverted) for worse.

Of singular significance to the hunting/antihunting argument is this: Shepard tells us that the diet, exercise, and social and spiritual norms distilled and instilled by natural selection in our hunter/gatherer forebears continue to be requisites for human health and happiness. Genetics confirms that as a species we've not had enough time in just ten thousand years of agriculture—and only half that of civilization (defined by literacy and urban living) to evolve one iota of change in our collective genome. And since modern urban living satisfies few of evolution's mandates for human health and happiness, there's your root of all dis-ease in the world today.

In sum, according to Paul Shepard: We are not living—physically, socially, psychologically, or spiritually—as we were designed to live. And that's the rub.

While hunting's critics often deride the activity as a barbaric anachronism—a filthy red remnant from our distant savage past—human ecology counters that since we evolved via hunting, and remain physically, mentally, and emotionally (genetically) exactly as we were then, to hunt is to *be* human. Thus, says Shepard, while hunting is indeed an anachronism in today's concrete world, it's a strongly positive anachronism that serves the invaluable purposes of continuing "to put leisure classes in close touch with nature, to provoke the study of natural history, and to nourish the idea of conservation."

And he's right. This truth is evidenced not only by the unfolding of my own hunter/conservationist worldview and many another of far greater significance, but by the remarkable growth of such hunter/conservation groups as Ducks Unlimited, Pheasants Forever, the Rocky Mountain Elk Foundation, and more.

Why do "modern" humans hunt? Because hunting is our genetic dictum, our generic heritage, its roots running as deep as humanity's tenure here on good green Earth. Deeper, in fact.

> O wonderful! O wonderful! O wonderful!
> I am food! I am food! I am food!
> I eat food! I eat food! I eat food!
> —*Taittiriya Upanishad*

The first hunters were the first humans. In fact, we were hunters long before we were fully human—assuming that "fully" is a title we still deserve today.

Mounting DNA evidence proclaims that our primordial ancestors split with the chimps to begin their long evolutionary trip "upward" five to seven million years ago. By four and a half million years ago, we'd achieved upright posture and moved out of the shadowy jungle onto the sunny savannas of ancient Africa . . . or, more precisely, we became creatures of the ecotones, those ecological edges where forest meets grassland.

Skipping forward, through another two million slow years of hominid evolution, the earliest prehuman believed to possess the uniquely human qualities of fashioning stone tools and using them to butcher large animal carcasses was *Australopithecus garhi.* He/she was not likely yet a hunter, but an avid scavenger, roaming the grasslands of Ethiopia at least two and a half million years ago, in search of fruits, vegetables, nuts . . . and easy meat.

Manufactured stone flakes have been found in conjunction with human remains even older than *A. garhi,* though not in clear association with butchered animal carcasses: So which came first—the dedicated scavenging of meat (muscle tissue), brains, and especially bone marrow, or the invention of tools to facilitate that scavenging?

If necessity is the mother of invention, then a specific need for manufactured stone cutting, chopping, and crushing tools must have preexisted their invention. Plucking and eating fruits, nuts, and grass seeds requires nothing fancier than fingers, or even just teeth, and root-digging technology need progress no farther than the end of a pointed stick.

"Man evolved as a hunter," proclaims the late/great Canadian biologist C. H. D. Clarke. "In South Africa, there were at one time [some two or three million years ago] two types of pre-men. One was a great shuffling hulk with a dentition that shows he was a vegetarian [*Australopithecus robustus*]. The other [*A. africanus*] was small and active, and fed on flesh as well as vegetable matter. This is the one that can be identified as having a place in the human pedigree. Vegetable gathering produced no tools, no forethought or planning, no tradition, no social organization. Pre-man the hunter, in developing and using all these for the chase, became man."

Furthermore, the brain is a high-metabolism organ and can grow—that is, evolve to greater size and complexity—only on a diet rich in the exact combination of fat-derived calories and nutrients found in meat—wild meat, that is. According to nutritionists, wild game such as elk, venison, and turkey contains five times the essential fatty acids found in domestic meat plus a nutritionally rich concentration of iron and other minerals, vitamins, and the protein imperative for brain development. (And all the while, wild meat delivers very low percentages of harmful fats compared to any domestic meat, red or white. The leanest domestic flesh is the white meat of turkeys, containing 3.2 grams of fat per 3.5-ounce serving. The same serving of elk or venison is a third leaner, with only 2.2 grams of fat.)

In a nut: Had our deep-time ancestors been vegetarians, we wouldn't be human today. In two direct and providential ways then—by offering an impetus for the development of lithic technology, or stone toolmaking, and by providing the right nutritional stuff for dynamic brain growth—meat eating presaged and facilitated our becoming human. The stage was now set for the third crucial contribution of carnivory to human evolution: hunting.

———

BY THE OPENING of the icy Pleistocene epoch, circa 1.6 million years ago, our prehuman ancestors had been scavenging meat and hunting catch-as-catch-could for thousands of generations, having evolved along the way from gregarious, bipedal brush apes to the penultimate hominid ("manlike") form, *Homo erectus*—just one short evolutionary leap from self-designated sapience and already armed with an arsenal of uniquely human intellectual, social, and technical skills.

By the close of the Pleistocene, ten to twelve thousand years ago, concurrent with the meltdown of the last continental glaciers, we were exactly as we are today—in every physical, mental, and genetic detail—and had been so for fifty to a hundred thousand, some say half a million, years.

Among the more important developments appearing during the Pleistocene was the refinement of verbal communication: spoken language. Complex language, rich in metaphor, reflects a mastery of abstract thought, which, in reciprocal turn, powers metaphorical language—Everyman's poetry—and facilitates, even compels, the definitively human quality we call spirituality.

And the pathway for all such development—Paul Shepard doesn't just tell us this but explains the process in carefully documented scientific detail—was our blood-bonded kinship, our daily give and take, with natural wildness, forged through millions of years—the final million of the Pleistocene especially—spent hunting, and being hunted by, large wild beasts. Shepard summarizes: "The dynamic of escape and pursuit is the great sculptor of brains. . . . Hunter and hunted are engaged in an upward, reciprocal spiral of consciousness with its constituents of stratagem and insight . . . the progressive refining of mind by cycles of predator and prey whose dances [through time and shared experience] became less and less random encounters, more and more choreographed."

And so it is, proclaims Dr. Shepard, stoutly backed by many notable others, that contemporary humanity's seemingly instinctive—we could say spiritual—need to hunt, although we no longer "need" to hunt in order to survive, along with all the richness of physical, mental, and aesthetic experience it implies, is in *fact* instinctive, arising from the ancient depths of the human genome. Similarly, an ongoing genetic need to be hunted is built into the elk, the deer, the antelope. This is Shepard's double-edged Sacred Game of give and take, without which nothing would be the same. (Without which nothing would *be*.)

Certainly, the greater an animal's reasoning abilities, the less its actions are dictated by knee-jerk instinct. Thus we human animals could, as most "civilized" folk today in fact do, forget our evolutionary history as hunters. Ignoring or even denying our heritage as predatory omnivores, most modern humans contract professional proxies to do their killing and cutting for them, politely out of sight (and thus out of mind). It's the easy way out; the ostrich's philosophy. Yet, by so doing, we are sanctioning the cultural reduction of what, before the agricultural "revolution," was viewed as sacred flesh—to lowly modern "product."

Worse yet, some among us affect moral superiority to the Sacred Game while denying responsibility for the daily deaths that feed us, eating only veggies, thus providing our omnivorous bodies and spirits just half a loaf while pretending to do no greater harm. Sorry, but it's just not so: No body rides for free.

I didn't fight my way to the top of the food chain to be a vegetarian! proclaims a locally popular bumper sticker.

If we aren't supposed to eat animals, why are they made of meat? chides another.

Cowboy philosophy, that, yet—just so.

I too could buy my meat at the supermarket or (re)turn to vegetarianism. But I don't and won't. Not so long as I have wilder, more natural options—which, tragically, most people in today's human-made world do not. I am lucky. And so long as there's a hunting or fishing season to enjoy, or roadkill to scavenge, I shall remain a forager. Notwithstanding its rejection by a minority, this is a basic, universal, human preference, evident in our appetites and physical design, confirmed by Paul Shepard whispering in our collective ear that "man is in part a carnivore . . . genetically programmed to pursue, attack, and kill large mammals for food. To the extent that men do not do so, they are not fully human."

In sum, I propose that to hunt, kill, and devour the flesh of creatures wild and free is not only the most natural possible exercise for body and spirit: it represents a palpable and significant, if only partial, return to our evolved animal heritage. Viewed in this light, honorable hunting is a spiritual sacrament, a humbling genuflection to our evolutionary design, genetic plan, and nutritional needs, as well as a sacred affirmation of our ancient blood-bond with the wildlings that for millions of years fed us, fed on us, and, in time, made us human. Thus were we created.

Obeying instinct, doing what comes natural, provides the bedrock motivation for all serious, thoughtful, ethical hunting today, transporting us back, spiritually if not temporally, to the fecund Pleistocene, to the lithic logic of sharpened stone, where it all began—where *we all* began. "In my imagination," wrote the inimitable desert philosopher Edward Abbey, "desire and love and death lead through the wilderness of human life into the wilderness of the natural world—round and round, perhaps forever, back again to wherever it is we began."

"I think we *Homo sapiens* lost something," writes Rob Schultheis in *Bone Games,* "some vital part of ourselves, when we gave up the hunting and gathering life for the fettered, programmed existence of agriculture, and later, industry. Consider just the moves: the lookout, the spoor, the stalk, the cross-country chase, the dead-on throw were replaced by the cramped, repetitive action of stoop labor, the planting stick and shovel, and worse, the quill, keyboard, and computer idiot stick. Time, which once rolled out in loose, measureless rounds of sun and moon, was chopped up into lengths and tied in knots; space, the free earth of the Paleolithic, was quartered and fenced: off limits; trespassers will be prostituted."

Thinking along parallel lines, another wise friend once commented: "In the descriptions of people who depended directly on the natural world for food and shelter and whose stories and rituals revealed an intimacy with the plants and animals sharing their lives, I find human traditions that offer a framework for my own feelings."

Paul Shepard refers to this same hopeful, optimistic, circular/cyclical view—consulting our deep past to guide our collective future—as "postmodern primitivism" or, as he titled one of his most important books, *Coming Home to the Pleistocene.*

And it's just such a future primitivism that's in play here and now in this weirdly wonderful Wild West place where I've come to hunt the wary, and wearying, pronghorn.

———

BLESSED EVENING ARRIVES at last, shoving long cool shadows across the moonscape valley below, signaling that the time is ripe, in fact beyond ripe, to put aside my campfire philosophy and attempt another guaranteed joyful (if predictably meatless) sneak on those fur-covered, fork-horned, bug-eyed mammalian rockets of the American West that we erroneously call "antelope." In fact, North America has no antelope. And the pronghorn, *Antelocapra americana,* which rather resembles a cross between a goat and a deer but is none of either—has no close relatives anywhere on earth.

For the past half-hour, a dozen prongies have been grazing carelessly in a grassy bowl a quarter-mile below me—courageously close by the species' notoriously standoffish standards. And just moments ago, a score or so more appeared on the southern horizon half a mile out—golden glowing dots in the distance, like so many pyrite nuggets, scattered beneath the late-day sun.

As I watch, the gathering animals feed fast as caribou toward the base of the long, narrow knoll where I'm sitting in plain sight, binoculars in hand. Seeing this swelling swarm of prongies, including several legal bucks, it's tempting to run one last fast stalk. But the day has grown too short; I should have gotten after it an hour ago. The odds, at this point, are far too low to justify the effort. And I'm simply too tired.

Been tired since this afternoon, when I sagged back to camp after an invigorating two-hour sneak-and-peek that took me to within thirty-five yards of a beautiful big buck with high heavy horns, a barrel chest, and thick meaty thighs. Following a fruitless morning spent sitting in would-be ambush over a tiny pool in an intermittent creek, and just after a rowdy midday thunderstorm, I spotted the animal lying beneath a rainbow—so far away that even through eight-power optics I could see no horns. Moreover, from that great distance the creature looked small and pale; probably a doe. Then the sun winked out briefly and a jagged line extending from horn-tip down around bulging eye and across bony cheek flashed black as ebony—a buck, thus legal prey.

After assessing the trend of the buck's stop-and-go grazing and contemplating the corrugated terrain along his route ahead, I contoured half a mile in that direction, then leaned into the climb, keeping to the bottom of a shallow arroyo. Every hundred yards or so I'd creep to the lip of the swale and glass until I relocated my potential target, adjusting my route accordingly.

The closer I got, the shallower the arroyo and the fewer the trees, until at last I was on a level with the unsuspecting prongy. Just two brushy little piñons remained between us. The penultimate tree—which I'd knee-crawled to without getting caught and was skulking behind—stood just fifty yards from the buck. If I could only sneak from there to the final pine and lean out around it without being busted, I'd have a twenty-yard shot, corresponding handily with my maximum dead-accurate stick-bow range. Peeking through parted limbs, I noted with relief that the buck had bedded, facing away.

From long habit, I flicked my Bic to check the breeze, even though some hunters claim pronghorns pay little attention to scent. Indeed, as my biologist buddy Tom Beck so eloquently states that case: "Your basic prongy is a whole 'nuther critter from your basic deer." And he's right, as we've seen; the two are related by neither taxonomy nor disposition. Yet I know from hard experience that pronghorns do listen to their noses, scent-spooking at times from hundreds of yards away. Besides, after a lifetime of hunting, it's a habit—reading and heeding the wind—that I'm helpless to ignore.

The lighter's little flame bent directly toward me; the breeze was in my favor. Thus refortified with hope, unto the breach I went, slow as any snail— out and around the sheltering piñon, heart booming like a timpani. Slow, slow, keeping always that last critical tree exactly between the prongy and me, taking one baby step at a time, placing each boot gently down, respectful of the crunchy volcanic rock underfoot . . .

Fifteen exhilarating, exhausting minutes passed as I closed the gap to within ten critical steps of the bushy little evergreen—which I, at least, could not see through. And just beyond lay my unsuspecting prey.

Even as I was permitting myself to think, *Good grief, man, you may actually make it this time*—my fingers already tightening on the bowstring and my mind's mouth watering in anticipation of grilled antelope loin steaks—

the bad news arrived. *Wheeeee!* The dread alarm-sneeze sounded. The buck was standing, rigid, staring arrows at me.

I became a statue, but too late. *Wheeeee!* With rump hairs flared and flashing electric white, he was going, going, and . . . I watched in awe as this most graceful of American mammals sailed like a great wingless bird over the nearest rise and disappeared into memory.

Oh to be a pronghorn—to run flat-out and never stop!

Since I'd come to expect no more, my disappointment was manageable. Like youth and other fleeting pleasures, it sure was fun while it lasted.

With the buck long gone, my feet hot, my legs aching from prolonged isometric strain, and no other game in sight, I glanced down at my watch, then up at the sun—sagging well past its apogee—and decided to pack it in for the day. Make camp early and kick back. But tomorrow . . .

It's a game we simply must play, we predators and our prey. To paraphrase Chief Sitting Bull: If hunters do not hunt, we will die of heartbreak—and so will the antelope. They need us, even as we need them. That's the way wild life's meant to be. That's the way it *has* to be. This is the heart of our sacred unifying wildness, the wily antelope and me.

Abbey said: "We are kindred all of us, killer and victim, predator and prey."

Shepard said: "Wildness is what I kill and eat, because I too am wild." I say: To hell with virtuality; let's get real again.

———

NOT SO LONG AGO, I was chatting with one of the participants in a nature writing workshop, which I was "facilitating" at the Yellowstone Institute. Bob was (and likely still is) a retired, octogenarian, Roman Catholic priest with a keen and admirable—borderline heretical—interest in "spirituality in nature." Bob's brand of spirituality, of course, was creations apart from my own, as he'd doubtless deciphered across our four days together. So when, inevitably, Father Bob inquired about the precise nature of my religious beliefs—and I wasn't feeling disposed, then and there, to dive into an involved explanation of my passion for "Earthiesm" (more about which later), being in something of a rush to get out and hobnob with the grizzlies and wolves—I said simply, "atheist."

"You mean 'agnostic,' don't you?" Father Bob rejoined politely, seeming truly concerned.

"No," I said. "I mean atheist. Or more precisely, "Earthiest," though we needn't get into the distinctions. Let's just say that for me, the meaning of life, the 'secret of the universe,' comes down to . . . biology."

"Biology," Bob mumbled, tasting the word in his mouth. After a moment he nodded politely and fell quiet. End of conversation.

This—the ultimate importance of biology in our world—is what Paul Shepard is trying to share with us, relating it to all aspects of human being: past, present, future; surreal as well as corporeal. As thoughtful nonhunter Brooke Williams, in his self-transformative *Halflives,* so succinctly condenses the foundation of Shepard's work and my own spirituality: "In primitive societies, hunting is embedded in custom, ritual, and story. Hunting forges deep obligations, people to people and people to the gods. It weaves a bond between the hunter and the wild landscape so pure and complete that the line where one ends and other begins is invisible."

Onward then: out of the shadows of our formative past, through the flickering firelight of life, into the foggy future. As the Spanish philosopher José Ortega y Gasset points out: "Like the hunter in the absolute *outside* of the countryside, the philosopher is the alert man in the absolute *inside* of ideas, which are also an unconquerable and dangerous jungle. As problematic a task as hunting, meditation always runs the risk of returning empty-handed. Hardly anyone can fail to know the probability of this result if he [or, of course, she] has tried, as I have . . . to hunt down the hunt."

So—tomorrow and tomorrow and . . . back to the hunt.

2

Dersu Uzala:
A Hunter/Conservationist
Paradigm

Only a personal relationship with Earth makes us love it.

—Leonardo Boff

LIKE SO MANY YOUNG MALES of my (Baby Blooper) generation, I grew up fantasizing myself into the frontier adventures of various literary heroes real and fictive—joining A. B. Guthrie's inimitable *Big Sky* mountain men, Lewis and Clark, James Willard Schultz (*My Life as an Indian*), Jim Corbett (*Man-Eaters of Kumaon*), and, transporting frontier romance into the twentieth century, those natty nabobs of manly sport, Teddy (Roosevelt) and Ernie (Hemingway).

Recently I discovered yet another literary wilderness hero, a special favorite because he, like me, is a hunter, a tree-hugger, and an unrepentant animist. I first met Dersu Uzala, Manchurian mountain man, in the 1975 Academy Award winning film bearing his name. Directed by Akira Kurosawa (the Japanese Francis Ford Coppola), *Dersu Uzala* was filmed on location in the remote Russian Far East. It was so good that I watched it twice, back to back.

When I learned that Kurosawa's script was based on the book *Dersu the Trapper,* by the Russian writer and explorer Vladimir Arseniev, I tracked it down and read it up. Although it's not "hunting literature" per se and has no out-front environmental message, *Dersu the Trapper* stands, nonetheless, as one of the finest literary examples of the hunter/conservationist extant.

———

DERSU UZALA WAS BORN in 1866 to a dwindling tribe of seminomadic Manchurian hunter/gatherers called Golds. On the occasion of their first meeting in 1902—when Dersu comes lumbering like a hungry bear from out of the taiga night and without a word of greeting makes himself at home at Arseniev's campfire—the Russian describes his uninvited guest as middle-aged and Mongolian in appearance: short and stocky with a "tremendous" chest and notably muscular throughout. His legs are "a trifle" bowed, his face weathered from a life lived under the sun, his broad smile fringed by a thin auburn mustache. But what impresses Arseniev most are Dersu's eyes: dark gray and calm, expressing "directness of character, good nature, and decision."

As this physique would suggest—and as Captain Arseniev soon learns—the "old" (in fact, middle-aged) Gold is a consummate woodsman:

> What to me seemed incomprehensible was to [Dersu] clear and simple. Sometimes he would pick up a spoor where with the best intention in the world I could see nothing. Yet he could tell that an old wapiti [elk] hind had passed that way with her fawn. They had browsed on *Spiraea* as they passed and then been suddenly startled at something and bolted. This was . . . done from lifelong habit, never to overlook details, and to be attentive to everything, and always observant. . . . Whenever I passed by a particularly obvious track, Dersu would chuckle, wag his head and say:
>
> "Hm! Just like small boy. Go walk, have eyes, no-can look-see, no-can understand. True, that sort man live in towns. No need go hunt wapiti. Want eat, go buy meat. No-can live all-alone in taiga, soon go-lost. . . . How you go many years in taiga, don't understand?"

In many ways, Dersu Uzala personifies the Noble Savage of romantic literature—a charismatic Caliban, a gentle wildman who lives by hunting wapiti for their hides and meat and trapping a few sable, which he barters for ammunition and other essential trade goods. His home is a small square of tent cloth, his bed a goatskin, his clothing mostly buckskin, and all of his worldly plunder rides easily in a handmade birch-bark pack. As Dersu explains to Arseniev in his quaint pidgin-Russian (just one of several languages the "primitive" speaks): "Me all time go hunt; no other work . . . only know hunt. Me no got house; me all time live moving; light fire, make tent, sleep; all time go hunt, how have house?"

DERSU'S CHRONICLER, Vladimir K. Arseniev, is an equally compelling figure. Despite his passion for word-painting himself as a greenhorn—a literary device, no doubt, intended to make Dersu's masterful woodsmanship shine all the brighter—Arseniev was in fact the leading Far East explorer of his day and a Russian national hero. Educated in a time and place where illiteracy ruled, Arseniev was a phenomenon: naturalist, surveyor, military officer, hunter, humanitarian, careful journalist, and master storyteller.

As a young man, his wanderlust having been honed by James Fenimore Cooper and other North American frontier literary dramatists, Arseniev managed to tour the American West while it still had a bit of its bark on. With the publication of *Dersu the Trapper* in 1928, Arseniev became the first Russian writer to inject an authentic Old West flavor into Old East literature.

Today *Dersu the Trapper* remains a classic in Russian and several other languages. Curiously, the 1930 English translation, by Malcolm Burr, published in America in 1941 by Dutton, enjoyed only brief popularity—blown away by the winds of war, I suppose. In 1996, *Dersu the Trapper* was republished in its full original text, including ink sketches by Arseniev.

What makes Dersu so compelling is Arseniev's unbiased presentation of the man as an animistic archetype—an anthropologically accurate, anachronistic example of how, for tens of thousands of generations, we all lived and thought, back when every human on earth was a professional hunter and gatherer. In the person of Dersu, this ancient foraging tradition, and the down-to-earth ecospiritualism it fosters and reflects, are brought forward into the opening decade of the twentieth century: rendered contemporary.

Although Dersu is Asian, he somehow speaks more directly, resonates more clearly, to the typical white, middle-aged, North American hunter's ear than do our own Native Americans, whose animism was long ago polluted by Christianity and who, for the most part, had adopted a horse-mounted economy before they began being quoted in print. While the Native American view of nature has often been romanticized in literature, culminating in the sickeningly sweet, New Age pap so common today—difficult for the contemporary hunter to identify with—Arseniev's portrayal of Dersu is refreshingly down to earth.

———

CENTRAL TO OUR CONCERN here is the fact that Dersu's worldview, representative of all hunting cultures past and present, even most horse-mounted Native Americans, is animistic. From the Latin *anima*, for soul, as in "animate" and "animal," animism is a zoomorphic (animal-oriented), deeply spiritual worldview that sees all of nature as sentient and significant. While only humans are human, all living things embody humanlike qualities, being possessed of feelings, pride, and a keen sense of bilateral reciprocity: Treat nature with benevolent respect, and nature will return that kindness—and vice versa of course.

Early in their acquaintance, Dersu's animistic tendency to humanize almost everything requires a bit of sorting out by his civilized Russian companion. After a successful wild boar hunt, for instance, Arseniev recalls:

"The [animal] killed by Dersu was a two-year-old sow. I asked why he had not shot a boar.

"'Him old man,' he explained, speaking of the huge tusker.

'Him bad to eat; meat much smell.'

"I was struck that Dersu spoke of the boar as a man, and asked him why he did so.

"'Him all same man, different shirt. Him know everything: know traps, know angry, know all 'round, all same man.'"

This democratizing of nature doesn't imply that animists "worship" forest "gods." Certainly, hunter/gatherers employ a plethora of customs, including ceremonies and symbols that may seem superstitious to outside observers, by way of acknowledging and honoring the spiritual essence of the natural world that feeds and clothes them. But all religions, animism to Zoroastrianism, incorporate ceremony and symbol. They not only arise out of supernaturalism but are steeped in it: religion *is* supernaturalism. Roman Catholicism provides a massively weighted example.

By comparison, animism, being earth-bound rather than sky-bound, and life-oriented rather than afterlife-oriented, is profoundly practical. Satirist Ambrose Bierce nailed it precisely when he described an animist as "a benighted creature who has the folly to worship something he can actually see and feel."

There's a humorous instance of animism in action in *Dersu the Trapper.* When Dersu and Arseniev discover they are being stalked by "Amba," the Gold name for the fearsome Siberian tiger, Dersu neither desires to kill the cat, as a westerner almost certainly would, nor falls to his knees and prays to the beast for mercy, as western prejudice would expect of the "savage mentality." Rather, the aborigine opts to negotiate with the cat. As Arseniev recalls the episode:

"Dersu stopped and, turning to the side where the tiger was hidden, called out loudly, in a voice in which could be heard a note of irritation: 'Why you go behind, Amba? What you want, Amba? We go our way, you go yours; you no bother us. Why you keep come behind? Taiga big place, room for us and you.'"

Having spoke his piece, Dersu turns his back on the tiger and walks confidently away. Amba, having apparently gotten the message, politely disappears.

To be sure, there are distinctly bad as well as good berings in the traditional hunter's universe. And most—like humans and tigers and, among American Indians, the coyote—are some of both. But good, bad, or changeling, all of nature is part of the animistic family, all is alive. Not only animals and plants, but wind, water, fire, ice, and place; all is invested with intelligence and spirit (that ineffable essence that today's conventional supernaturalists call "soul"); everything is connected, everything is important, nothing is impotent. Nothing, in sum, is to be taken for granted or willfully abused.

Slowly, by lesson and example, Arseniev comes to comprehend and respect Dersu's self-humbling, ecocentric, animistic worldview. One night, for example, as the teapot rattles annoyingly on the campfire, Dersu, feigning anger, accuses the fire of being "naughty" and calls the boiling water a "bad man." Puzzled anew, Arseniev asks how fire and water, not being alive, can be "men." Dersu is incredulous. He's amazed that anyone, "even a Russian," could fail to see something so obvious: "Water," he patiently explains, "him can cry, him can sing, him can play." Then, pointing at the leaping, dancing fire, he adds, "Him too all same man." In Dersu's view—the animistic view, the archetypal hunter's view—water and fire *are* alive.

Intelligent and objective, Arseniev openly celebrates the pragmatic beauty of Dersu's democratically animistic worldview, noting that "this primitive soul saw in water a living force; he saw its quiet flow and heard its roar in time of flood."

When Arseniev remarks one morning on the beauty of the sunrise, Dersu confirms his friend's praise by replying: "Him [the sun] most important man. . . . If he go, all other things go."

"In those simple words," Arseniev observes, "there was pure animism, but also profound philosophy."

———

PRECIOUS FEW pure hunter/gatherers and nonpolluted animists survive in today's overcrowded world. Their kind has largely been co-opted, "reserved," or exterminated by the mercenary march of agricultural/industrial civilization. Even at the opening of the twentieth century, even in remote Manchuria, Dersu Uzala is among the final untainted representatives of this dying breed to whom—among other "primitive" qualities largely lost to civilized humanity—altruism comes natural.

For example: When Dersu and Arseniev happen upon a deserted trapper's shack, they clean and repair it, then bivouac inside for the night. Next morning, as they're preparing to leave, Dersu restocks the shanty with firewood, then wraps a few matches, a pinch of salt, and some rice in protective birch-bark bundles, which he hangs inside. When Arseniev asks his enigmatic friend if he's provisioning the place because he plans to return soon, Dersu says no. "Some other man he come . . . he find dry wood, he find matches, he find food, not die."

Arseniev finds such generalized generosity profoundly moving. He's amazed that his barbaric buddy "should bother his head about an unknown man whom he never would see, and who would never know who had left him the provisions. . . . Why is it that among town-dwellers this forethought for the interests of others has completely disappeared, though no doubt it was once there?"

Vladimir, my ingenuous friend! How you go so long in towns and no-can-see? Cities create crowding, crowding promotes estrangement, strangers spread fear, and fear breeds contempt.

In happy contrast, in remote bush areas of Alaska and Canada even today—and quite likely in rural Manchuria and Siberia as well—the "savage" tradition of leaving transient bivouac cabins unlocked and stocked with survival goods remains alive and well.

―――――

SUBTLY, as their years and adventures together mount, the sophisticated and westernized Arseniev increasingly folds aspects of Dersu's "simple" worldview into his own—confirming that going native comes natural in natural settings. Increasingly, Arseniev utters such shamelessly animistic sentiments as, "Every time I enter a forest which stretches for several miles, I have an involuntary feeling of something like humility. Such a primitive forest is a kind of elemental force."

Yes, comrade, *it is* a force. It's in our blood, in fact, our very heart's blood. Or, to employ genetic rather than spiritual terminology, it's part of the human genome—the humility, awe, and visceral excitement that those whose senses haven't entirely been blunted by concrete, smog, and cacophony yet feel, experience, thrill to, in the presence of great unbroken tracts of true wilderness and the gloriously untenable creatures that animate such places.

And for good reason: Wildness is not only where, but to *what end*, we evolved. It is the great green stage on which the human drama auditioned, premiered, and played for all but the most recent and possibly final act of our existence. So wholly are we products of wild nature that when the last true wilderness is gone, we not only will cease to be hunters—that is, cease to be naturally wild in accordance with our natural design—but will be utterly lost as a species. A world bereft of wild places and wild creatures—a wholly humanized, civilized, and man-ipulated earth—will be far too bleak and lonely a place for sanity to survive. And then what?

This is why I say, every day and every way, to my fellow hunters as well as to hunting's critics: There is far, far more to the hunt than trophies, than recreation, or even than meat. The hunt is, or should be, a quiet, deeply personal rite, an active sacrament that reconnects us to our human/humane roots and realigns us with wild (being the only true) nature, our one and only home: past, present, forever. In this lesson and so many more, Dersu Uzala remains a uniquely wise teacher—to us as he was to Arseniev.

Unlike the Chinese, Korean, and Russian exploiters invading his beloved taiga—butchering forest and wildlife wholesale—Dersu clings tenaciously to the ancient woodsman's code forbidding hoarding in any form, including coin, as well as the greed-driven taking of life.

Nor does Dersu kill for masochistic thrills or allow others to do so in his presence. When one of Arseniev's Cossack escorts raises his rifle for "target practice" at a sea lion swimming near a remote beach, Dersu pushes the man's rifle aside and gently counsels, "No need shoot. . . . No can take body; shoot all same wrong."

At this juncture, yet again, Arseniev is forced to question his own "advanced" culture's destructive, arrogant attitude toward nature. He asks rhetorically: "Why is it that we Europeans so often abuse our power and our weapons, and take the lives of animals thoughtlessly, as it were for mere amusement?"

Likewise I ask: Why too we Americans, hunters and non?

In response, some hunters will counter: "Well, why not?" And egging them on in this piggishness is a greedy and increasingly immoral hunting media.

———

ALDO LEOPOLD once quipped that "conservation" is what one thinks while chopping down a tree. Something similar can be said of the public perception of hunting: For most people, as important as what hunters do is *why* we do it. What are our motivations?

In this regard, arguably the most harmful force shaping both hunter behavior and public opinion of hunting today is the outdoor media—TV, radio, videos, and especially magazines. These public voices should be the primary tools for motivating ethical hunter behavior and, through positive moral leadership, in winning public support for ethical hunting. But overwhelmingly and increasingly, it's just the opposite.

"If you're a deer," proclaims the cover of a recent issue of a major commercial hunting magazine (can't recall which one, nor does it matter, as they're all of a lot), its pages laden with ads for products and services openly designed to help avoid the hard work and necessary skills of true hunting. *"If you're a deer, you're dead!"*

Contrast this "war on wildlife" theme, so common in modern hunting journalism, with the running message in Arseniev's *Dersu the Trapper.*

Since I'm most familiar with magazines, having worked with and for them for a quarter-century, and since magazines comprise the longest, strongest, and most visible arm of the outdoor media, I'll concentrate my criticisms in that realm. The problems with magazines—outdoor and otherwise—arise from the bottom line. Commercial periodicals derive the bulk of their income not from subscription or newsstand sales, but from advertising revenue. And in order to attract ads, especially the big glossy juicy ones, you need a strong reader base. Consequently and ironically, magazines often go so far as to offer subscriptions at a loss in order to attract more readers to use as bait to attract more (and more profitable) ads.

Owing to this influential relationship—with the outdoor industry paying the outdoor media's bills and providing the lion's share of its profits via advertising—the outdoor industry's control of magazine content and politics is ubiquitous and increasingly hurtful to hunting. At its worst, this influence is overt, leading publishers, editors, and writers around like lapdogs on short leashes, resulting in magazines that read and look like mail-order catalogs.

Even at its best, this negative influence is barely covert. For instance: Freelance writers quickly learn that it's fruitless to submit articles that are even glancingly critical of major advertisers or product categories—ATVs (all-terrain vehicles), for a particularly noisome example. After a while, you quit even trying.

Conversely, commercial outdoor magazines love to praise what they hawk. Sticking with the example of ATVs, many hook-and-bullet publications run regular "off-road" columns, features, and even special issues touting the latest in vehicles and accessories, thus endorsing and glorifying the bucket-butt slobbishness that motorized hunting promotes—all in hopes of attracting more ads for same. And it works. To hell with ethics.

And who in this market-driven world would even *consider* that hunters have a sacred duty to respect and preserve wild nature? Despite the obvious truth that ATVs damage ecosystems, disadvantage prey, debase those who use them, and infuriate those who are forced to see, hear, and smell them—ATV ads are highly profitable. Along with building a negative conservation paradigm, by pushing ATVs and other technogadgetry the media are encouraging hunters to hunt with their butts and wallets—cruising roads and trails, searching for easy targets—rather than their boots, brains, and hearts.

Another way the outdoor media betray hunters is by catering to the demands of loud minorities, such as the rich, powerful, and well-organized ATV/ORV (off-road vehicle) owner groups and extreme "hunters' rights" outfits—some of whom, industry-funded and headed by high-ticket professional flacks, generate stacks of letters to editors and politicians and wildlife management agencies on a plethora of "gimme" issues ("Gimme more of this! Gimme more of that!"), many of which are blatantly antinature.

By buckling to such political pressures, the hunting media become profit-followers rather than ethical leaders, willingly duped into believing that a mouthy minority of myopic hunters represents the mute majority, then doing all they can to popularize that self-serving myth.

Another example of negative influence on hunters by the commercial outdoor media is the lack in those pages of thoughtful articles about hunting tradition, ethics, and conservation. And any hint of spiritualism, of course, is greeted with cries of "voodoo nonsense." The reason for this shortage of the most essential topics, quite simply, is that tradition, ethics, conservation, and spiritual "voodoo" don't sell magazines, ads, or products.

Nor are the outdoor media entirely to blame. Today most magazine-reading hunters crave, and demand, such "service" features as how-to, where-to, and product evaluations, thus aiding and abetting this incestuous ethical degeneration. A positive step would be for outdoor writers and editors to start crafting their "service" articles to include subtle ethics, conservation, and spiritual themes—in effect, sandwiching truth and reality between opening and closing slices of the entertainment and information many modern hunters seem to want.

Certainly there are exceptions to all the above. And foremost among them is *Bugle*—the big, bright, quarter-million-circulation publication of the Rocky Mountain Elk Foundation (RMEF), arguably America's most enthusiastic and effective hunter/conservation group. At once fun, informative, reverent, ethical, literate, and handsome, *Bugle,* though not a major profit generator, nonetheless pays its own way. Unfortunately, *Bugle* leads only a tiny number of contemporary "sporting" publications—among them *Mule Deer* and *Pheasants Forever* (both published by hunter/conservation groups)—that V. K. Arseniev would likely have been proud to write for.

Fortunately, the hard-jacketed hunting media—that is, books—being free from advertiser influence and (generally) aimed at a more intelligent and discerning readership, is making a somewhat better show of itself, especially in recent years. But the muck is heaped high in some corners of this stable as well.

(Nor, to give fair time, is the news any better from the "animal rights" media, who are every bit as disingenuous as commercial hook-and-bullet rags, although for utterly different motives. For one example, which should not be an example at all, I can cite *E, The Environmental Magazine.* While not published by an animal rights group, *E* is (or was, last time I read it) overtly antihunting—to the ridiculous and censorial point of refusing (last time I inquired) to publish any article about hunting written by a hunter, no matter how critical that article might be of hunting.)

Past or present, I can think of no finer literary portrayal of hunting as it so long was and must again become, than *Dersu the Trapper:* a literary animist/hunter/conservationist paradigm. In the person of Dersu Uzala, we see the living embodiment of the "personal relationship with earth" that ecophilosopher Leonardo Boff cautions is necessary to "make us love it."

In order to assure a meaningful hunting future for our children and theirs, in order to give something in return for the many joyful returns hunting and wild nature have given to us, in order to demonstrate to the non-hunting majority that hunting is worth preserving, in order to glean the best the hunt has to offer—we who call ourselves hunters must relearn our ancient ancestors' respect for *all* of nature, not only that which serves us directly. We must learn to cool our culturally codified impulse to control and modify and meddle. And most of all, we must join hands and hearts with likeminded nonhunters to celebrate and preserve wildness for its own sake.

In the end, we must prove, and improve, our hunter's love for earth, like Dersu, by striving ever more diligently to "take every kind of care" of what remains of truly wild nature, inside of us as well as out.

3

Hunting Down
the Hunt: Meditations
on José Ortega y Gasset

Like the hunter in the absolute *outside* of the countryside, the philosopher is the alert man in the absolute *inside* of ideas.

—José Ortega y Gasset

PERHAPS THE MOST-QUOTED LUMINARY in hunting "literature" today is the Spanish philosopher, educator, and essayist José Ortega y Gasset. Here's a triplet of oft-sung examples: "Every good hunter is uneasy in the depths of his conscience when faced with the death he is about to inflict on the enchanting animal. . . . The beauty of hunting lies in the fact that it is always problematic. . . . The inequality between hunter and hunted should not be allowed to become excessive."

In fact, I'd say Ortega is quoted too often by "outdoor communicators" these days, placing his wisdom at risk of becoming clichéd. But then I'm hardly in a position to complain, having once or twice, maybe thrice, borrowed Ortega's words myself. It's hard not to—for Ortega's classic little tome, *Meditations on Hunting,* distills the essence of the true hunter's heart and makes you want to exclaim: "That's *it!* Ortega says perfectly what I've always felt but could never put into words." Since his name, and words, appear scattered throughout the coming pages, it's worth asking right now: Who was this "famous" yet little-known thinker?

———

JOSÉ ORTEGA Y GASSET has been called "one of the greatest Spaniards of all time." Born in Madrid in 1883, he died there in 1955. In his youth he earned a doctorate in philosophy from the University of Madrid, where he went on to enjoy a long and distinguished career as an educator.

In his high-profile life as an academic, intellectual, and political leader, Ortega functioned as an aristocrat in an acutely class-conscious culture. In such circumstances, lesser men invariably succumb to elitism and snobbery. Yet, as a moral adviser to the Spanish Republican government and a member of its parliament, Ortega disdained class arrogance, speaking out courageously against totalitarianism. (In *Meditations,* for instance, he complains that "one of the causes of the French Revolution was the irritation the country people felt because they were not allowed to hunt.")

From the outbreak of the Spanish Civil War to the end of World War II—a violent interlude during which liberal intellectuals like Ortega were neither welcome nor safe at home—the outspoken academic lived in serial exile in Germany, France, Holland, Argentina, and Portugal and conducted lecture tours in Switzerland, the United States, and elsewhere. Four years before his death, the grand old gentleman of letters was awarded an honorary doctorate by the University of Glasgow.

———

AS A PHILOSOPHER, Ortega is something of an anomaly, praised by some, dismissed by others. As translator Phillip W. Silver summarizes the uncharitable view, Ortega was "an unclassifiable, vaguely philosophical essayist [who] seemed to stand virtually alone, with no context to support or explain him." Rather than adhering dogmatically to any established school of thought, Ortega had an alert, hungry mind—a hunter's mind, if you will—that sampled a broad smorgasbord of ideas. Such "revolutionary" freethinking, of course, is typically feared, abhorred, and criticized by true believers of every stripe. The provocative titles of Ortega's books suggest the sprawl of his intellectual interests: *The Revolt of the Masses, Invertebrate*

Spain, Toward a Philosophy of History, Concord and Liberty, Man and Crises, On Love, What Is Philosophy?, Meditations on Quixote, The Dehumanization of Art, and more.

If I were asked to classify Ortega's scattered philosophy, I'd go with "existential phenomenologist," given his bent for chasing after such slippery questions as: How should we deal with the reality of life? As rational animals, how should we live and where should we look for guidance? Ortega's answers generally boil down to a pragmatic (and currently popular) maxim: "I am I and my surroundings."

In addition to phenomenology (the study of consciousness in general and self-awareness in particular), Ortega's other great intellectual passions included aesthetics and art. And in hunting he saw all three at play. As Howard B. Wescott, who translated and introduced *Meditations on Hunting,* points out: "It is not surprising . . . that the drama of the hunt should interest a man whose philosophical stance is based on the drama of biographical life."

But many have written passionately about the drama of the hunt. What is it about Ortega that makes him so special, so keenly quotable, so tenaciously popular? Does he really deserve the unprecedented ink and adulation he's receiving in the sporting press of late? Or is he just another pretty writer full of sound and fury etc. etc.? More important, can we who live and hunt in America today safely rely on this early twentieth-century European's ideas as valid to our concerns—as worth repeating, defending, even living by?

———

ONE COMMON CRITICISM of Ortega is that, under close scrutiny, some of his pithiest pronouncements boil down to mere platitudes, moral cop-outs, intellectual throwaways. And frankly, when plucked like cherries from the tree of context, some of Ortega's aphorisms do resonate as hollow. For example: By way of launching this discussion, I listed three quotes, hoping you'd have encountered at least one of them before, whether or not you recall Ortega as the author. Number three in my initial list was this: "The only adequate response to a being that lives obsessed with avoiding capture is to try to catch it."

To experienced hunters, evolutionists, and wildlife biologists, Ortega's meaning is clear: Elk evolved as prey, while humans evolved as part-time prey and full-time predators; so what could be more natural than for our human hunting instincts to be excited by the deer's keenly evasive instincts, which can, in this context, be interpreted as a tease? Bears, wolves, lions, and most other mammalian predators are instinctively inclined to pursue anything that runs away. Which is to say: The exceedingly problematic challenge of "catching" the elusive deer helps to explain why we so love to hunt them.

But to a nonhunter/evolutionist/biologist, that quote, lifted from the context that shapes and explains it, held aloft like a fish out of water, sounds, in the words of one critic, "silly and nonsensical." And, similarly abused, so do others. But place Ortega's "silly and nonsensical" cherries back in the tree of context from which they were carelessly plucked by some overreaching "outdoor communicator"—say, the widely scorned "one does not hunt in order to kill; to the contrary, one kills in order to have hunted"—and see what we get:

> In utilitarian hunting the true purpose of the hunter, what he seeks and values, is the death of the animal. Everything else that he does before that is merely a means for achieving that end, which is its formal purpose. But in hunting as a sport this order of means to end is reversed. . . . What interests [the sportsman] is everything that he had to do to achieve that death—that is, the hunt. Therefore what was before only a means to an end is now an end in itself. Death is essential because without it there is no authentic hunting: the killing of the animal is the natural end of the hunt and the goal of hunting itself, not of the hunter. . . . To sum up, one does not hunt in order to kill; on the contrary, one kills in order to have hunted. If one were to present the sportsman with the death of the animal as a gift, he would refuse it. What he is after is having to win it.

As translator Wescott points out, Ortega examines hunting "from the perspective of what that activity has demanded of a particular individual and what it has meant to him within the context of his life." And this, to my way of thinking, is the heart of the matter.

Taken as a whole, Ortega's *Meditations* is a compelling call for ethical hunting. Hunting without significant effort, skill, prey compassion, conservation ethic, tenacity, humility, sense of awe in nature, and personal honor—all motivated by a deep-seeded need (spiritual if not physical) for the best that authentic experience has to offer—while it may be pursuing and it may be killing, is not hunting.

——————

IRONICALLY, *Meditations on Hunting*—Ortega's best-known work today in America if not worldwide—came about almost by accident. It was written, in 1942, not as a book, but as a prologue to the sporting memoirs of one Edward, Count Yerbes. While the count's *Twenty Years a Big-Game Hunter* has been largely forgotten for fifty years, across that same half-century Ortega's prologue, repackaged as a slender book, has remained continuously in print, grown steadily in popularity, and won translation into several languages. Here in America, the indefatigable Paul Shepard was so impressed with the Spanish edition of *Meditations* that he arranged to have it translated and published in English.

So brief is *Meditations* that even a plodding reader can consume it, if not fully comprehend it, in one devoted evening. Stripped of all fluff and padding, *Meditations* comprises more or less seventy-five pages of text (depending on the edition), divided into ten chapters of varying brevity.

I have neither the desire nor the credentials to critically dissect Ortega's thoughts page by page. Nor, I suspect, do you wish to be subjected here to so thorough an analysis. Yet to help balance the copious praise I'm otherwise heaping on Ortega, I will at least venture that the chapter "Polybius and Scipio Aemilianus" can be skipped over with no fear of missing anything of import and the chapter "The Scarcity of Game" is a study in redundancy and a Sisyphian belaboring of a single self-evident point.

Yet even "The Scarcity of Game" contains sparkling gems of wisdom. Particularly apropos now, in this age of gadget addiction and Pavlovian consumerism, is Ortega's caution against "confusing the hunt itself with that which merely *has to do* with [the hunt]." (And here I must wonder: Had Ortega read Thoreau, who, a century before, wrote that "our inventions are wont to be pretty toys which distract our attention from serious things"?)

Here also, Ortega is admirable in reminding us that "a very accomplished hunter should consider the supreme form of hunting that in which, alone in the mountains, he is at the same time the person who discovers the prey, the one who pursues it, and the one who fells it."

And, I hasten to add, the one who guts it, packs it home, butchers and eats it.

———

SUCH PICAYUNE CRITICISMS aside, the thoughts expressed in *Meditations on Hunting* are, overall, solid as granite. The reason Ortega is so often quoted is that he's so often quotable. This, in turn, is less because his prose is so "pretty" and more because his observations are so meaningful. What veteran of the chase could fail to beam with recognition upon hearing Ortega say:

"The hunter knows that he does not know what is going to happen, and this is one of the greatest attractions of his occupation. Thus he needs to prepare an attention . . . which does not consist in riveting itself on the presumed but consists precisely in not presuming anything and in avoiding inattentiveness. . . . The hunter is the alert man."

As a bonus, Ortega even slips in a few sly bits of esoteric humor: "The only people to have felt that they had a clear idea about [humanity's relationship to] the animals were the Cartesians. The truth is that they believed they had a clear idea about everything." Cartesians, also known as Cartesian dualists, were disciples of a cruel-hearted, Christian fundamentalist ("the world was made for man to use and abuse at whim") philosophy espoused by the French mathematician and writer, René Descartes (1596-1650).

Similarly, today, Ortega's *Meditations* offers powerful rebuttal to such pronouncements as erstwhile "animal rights" champion Cleveland Amory's charge that "hunting is an antiquated expression of macho self-aggrandizement, with no place in a civilized society."

To the contrary, Ortega argues, hunting (and the preservation of wildness it implies) is in fact essential to the maintenance of any truly "civilized society" and thus can never become obsolete. How can we expect to nurture our continued evolution as a species by ripping out ninety-nine percent of the roots of that evolution? Hunting and gathering, millions of years deep, are those roots. Ortega summarizes: "Man's being consisted first in being a hunter. . . . Pushed by reason, man is condemned to make progress, and this means that he is condemned to go farther and farther away from Nature, to construct in its place an artificial Nature." In other words: "virtual reality" foretold.

Honorable hunting—that vital "alert man" connection to humanity's true (evolved) essence—remains a strong element of individual and cultural sanity: an instinctive, thus joyfully "unreasonable," element that long disuse and cultural corrosion have weakened or even destroyed in many modern humans. Today's growing chorus of well-intended but biologically (thus philosophically) befuddled animal rights champions are proud-walking, loud-talking examples of this tragic loss.

Reading Ortega helps to repair these broken links. Take, for one telling instance, this noble admission from Paul Gruchow's enchanting *Boundary Waters:*

> The Spanish philosopher José Ortega y Gasset calls hunting a "vacation from the human condition." It is only when animal and human are engaged as hunter and hunted, he argues, that we cease to be fugitive from nature and confront it on its own terms. He condemns people like me who have substituted looking for hunting. We are, he says, voyeurs—still hunters, but only idealistically, platonically; we are guilty—the ultimate sin in his view—of affected piety. I have thought about that argument a lot. I want to refute it, but I confess that I am unable to say where the error lies.

Not all readers are so alert and charitable as Gruchow. Those with opposing philosophies will persist in finding fault with *Meditations*—as Ortega himself would do, no doubt, were he kicking yet today. After all, intellectual growth through self-education, introspection, and open-minded polemics is the name of the sophist's game; philosophy is the love of pursuing wisdom, not the love of being right.

Forward, then, in that pertinent pursuit.

PART II

Worldviews in Conflict

People carrying signs
Mostly saying "Hooray for our side!"

—Stephen Stills

4

Hunters: A Scientific Profile (with Personal Conclusions)

> What can be said about hunting that hasn't been said before? Hunting is one of the hardest things even to think about. Such a storm of conflicting emotions!
>
> —Edward Abbey

LET'S EXPAND A LITTLE on Abbey's exasperated query: What can be said, with objective authority, about hunters and their philosophical opposites comprising the animal rights movement—as individuals, as groups, and as subcultures? Who are we, and who are they? What motivates us, and what drives them? And most significantly, how did an "us versus them" conflict evolve in people's feelings about hunting in the first place?

A quarter-century ago, behavioral scientist Stephen R. Kellert—Yale University research professor and coeditor, with Edward O. Wilson, of *The Biophilia Hypothesis*—undertook to answer all these questions and more. Kellert's years-long study involved some thirty-two hundred participants, employed "personally conducted, in-depth, largely unstructured interviews," and included "a national investigation involving a highly structured, close-ended, forty-five-minute questionnaire." Kellert and his associates also followed hunters into the field for observation. In 1978, the Kellert study's conclusions were published in a fifty-eight-page report titled "Attitudes and Characteristics of Hunters and Antihunters and Related Policy Suggestions."

This chapter and the next revolve around that historic report, which I sometimes augment by quoting Kellert's more recent research and writings. (You'll find them all in the bibliography.) And, naturally, I've inserted generous doses of my own wholly unscientific observations, insights, and experience-informed opinions. Although Kellert writes in the past tense, I can attest that his findings, with few and minor exceptions, remain as valid now, at the millennium, as they were in the golden 1970s.

In the introduction to his report, Kellert explains his root interest in the hunting controversy. "More than any other subject," he notes, "views for and against hunting provided a kind of barometer for assessing people's much broader understandings of the natural world." And understanding these "understandings"—people's views about nature in general and wildlife in particular—is Stephen Kellert's academic forte, for which he is internationally recognized.

———

KELLERT BEGINS by looking at hunters. After offering the caution that for most purposes it's "simplistic and even foolish to discuss hunters as a whole," Kellert nonetheless contrasts some basic attitudes of hunters against those of nonhunters (including antihunters):

> One of the most striking attitude findings was the much greater naturalistic attitude of hunters compared to nonhunters. Hunters generally indicated more interest in the outdoors, and more affection and desire for contact with wildlife. Hunters also reported significantly greater interest in wilderness areas, in living in proximity to wildlife, and in backpacking, camping out, and fishing activities than nonhunters. . . . Thus, despite the somewhat paradoxical fact that hunters killed wild animals . . . they were characterized by substantially greater affection, interest, and lack of fear of animals than nonhunters.

Heeding his own caution, that's about as far as Kellert ventures in painting all hunters with one broad brush. Thereafter, he divides hunters into three broad groups according to their attitudes toward nature and wildlife: utili-tarian/meat hunters, dominionistic/sport hunters, and naturalistic/nature hunters. While the borders are blurred, exceptions abound, and statistics sometimes lie, my own experience suggests that Kellert's categories, as well as the data he uses to support them, are not only sound but revealing.

———

THE MOST STATISTICALLY SIGNIFICANT segment of Stephen Kellert's nation-wide sample—forty-four percent of those who reported having hunted within the past five years—claimed to hunt primarily for food. These utili-tarian/meat hunters tended to be older, lower than the national average in education and income, and to hail from rural, agricultural backgrounds. Although they tested fairly well on Kellert's animal-knowledge scale, meat hunters evidenced dominantly utilitarian attitudes toward animals—as reflected in their support of trapping, predator control, and other such "practical" uses and manipulation of wildlife.

More subtly, meat hunters expressed their dualistic attitudes toward wildlife through language, employing such agricultural terms as "crop," "cul-tivate," "harvest," and "renewable resource." (These same euphemisms have since infiltrated both outdoor journalism and wildlife management, insidi-ously reducing such lovely and lively creatures as deer, elk, and grouse to the level of turnips.)

But even among these matter-of-fact, meat-hungry carnivores, there's clearly more going on than just grocery shopping. Without belittling their fondness for lean, clean, delicious wild meat, I suspect that part of what most self-avowed meat hunters are "hunting" is the atavistic satisfaction of sea-sonally (and in many ways even ceremonially) returning to nature to per-sonally secure a token portion of food the good old-fashioned hard way. By so doing, ironically, they seem to be ritually circumventing the selfsame agri-culture they otherwise represent.

Or perhaps not: Kellert refers to an incipient "pioneering spirit" among meat hunters, noting their "special respect . . . for resourcefulness, self-reliance, and individuality." Certainly there exists a brief ecotonal phase in the transition from seminomadic foraging to sedentary agriculture during which small-scale subsistence farming/herding ("homesteading") is augmented with peripheral hunting and gathering in the surrounding, not-yet-denuded coun-tryside. Possibly, as Kellert suggests, this recent "pioneering" heritage, rather than any deep-time foraging instinct, is what modern meat hunters are attempting to honor and at least symbolically recapture via the hunt.

———

KELLERT'S SECOND CATEGORY—dominionistic/sport hunters—accounted for thirty-nine percent of those who reported having hunted within the past five years. Most of this lot were urban, had served in the military, and knew or cared little about nature. To this group, says Kellert, "the hunted animal was valued largely for the opportunities it provided to engage in a sporting activity involving mastery, competition, shooting skill, and expressions of prowess."

Indeed, for years I've been troubled by the fact that so many hunters can quote the ballistic intricacies of their weapons by heart, but know nothing about the animals those weapons are aimed at. For this bunch, the modern means of hunting have overwhelmed its ancient ends.

Along with their ignorance of nature, many of Kellert's "sport" hunters indicated a generalized indifference toward animals or even feared them. "This relative lack of affection for animals," says Kellert, "was indicated by responses to specific questions dealing with the perception of most wildlife as dangerous . . . reporting less desire for close contact with animals, supporting a limitation of undesirable animals . . . and reporting substantially less childhood and adult interest in animals."

Perhaps—and this is my ten-cent psychoanalysis, not Dr. Kellert's— such a detached, even fearful, attitude provides a rationale for the care-less killing in which the worst of "sport" hunters so gleefully indulge. This is the "dangerous game" mindset to whom some commercial outdoor publications openly pander through continual replay—in cover art, headlines, photos, articles, and political opinions. Consider these recent examples from America's biggest hook-and-bullet periodical, *Outdoor Life:* "Shark Attacks" and "Mauled by a Grizzly" (February 1998); "Attack of the Killer Elk!" (September 1998); "Attacked by a Cougar," "O'Conner Hunts a Man-Eater," and "Killer Coyotes" (February 1999). And similarly, from *Petersen's Hunting* (no relation), "Blood Thirsty Kodiak Bears!" (November 1999). It follows (we are left to assume) that if wild animals are dangerous, they're de facto "bad," and killing them for no better reason than to collect their heads is not only justifiable but noble.

It's hard to sympathize with the nature-ignorant "Whack 'em and stack 'em!" mentality of many (though I don't think most) of Kellert's 1970s dominionistic/sport hunters—which mindset, brought forward to the turn of the millennium, includes the current epidemic in North America of what Aldo Leopold, half a century ago, dubbed "gadgeteers." Understandably, the dominionistic/sport is the type, among Kellert's trio of types, most despised by the animal rights/antihunter minority (frequently referred to by hunters simply as "the antis").

I share the antis' disgust with such moral abominations as egomaniacal trophy hunting, where acquiring antlers, horns, or skulls large enough to qualify for entry in any of several trophy record books (maintained by trophy-hunting "clubs") becomes the sole goal of the hunter. Just as disgusting are baiting (any baiting of any species); canned "hunting" (the execution-style killing of fenced-in wildlife); cheater technology (two-way radios that allow hunters to gang up on prey; night-vision optics, heat-sensing devices, and hearing enhancers, all of which give hunters unfair game-detection advantages; off-road vehicle abuse (providing lazy hunters easy access to distant game, often coupled with the motorized harassment of wildlife); public arrogance, bloodlust, and bluster; and so much more.

At the same time, I resent the antis' disingenuous efforts to taint all hunters with the stain of the lowest stratum of the dominionistic subset. By so doing, hunting's critics are diluting what would otherwise be strong moral arguments against specific behavior by, in Kellert's words, "simplistically and even foolishly discussing hunters as a whole."

———

OF COURSE, the antis are hardly alone in their fondness for misleading generalizations, half-truths, and flat-out lies. As *National Geographic* editor John G. Mitchell points out: "In the matter of hunting or not hunting, everyone is so desperately eager to be considered correct. Yet there is such an overburden of dishonesty on both sides."

And that dishonesty, on both sides, is prompted in more or less equal parts by ignorance, emotionalism, self-service—and, yes, heartfelt concern for what each side perceives as "right." It was Teddy Roosevelt who coined the term "lunatic fringe," predicting that the radical minority would always be rejected by mainstream Americans. With respect to hunting—even though the extremes grab the lion's share of media attention—there clearly remains a moderating majority of good sense and sanity rooted firmly in the middle. I'm referring to those selective critics of hunting who are not "anti" in any doctrinal, neo-Jainist sense, but are merely outraged by some of what they see marching beneath the guidon of hunting today. To such constructive critics I say: Yes, I share your outrage—and so do a growing and increasingly outspoken number of ethical hunters. As *Field & Stream* columnist George Reiger notes, if hunters someday lose their self-proclaimed "right" to hunt, it will come about as a direct result of being "too demanding of rights and too indifferent to responsibilities."

The dominionistic/sport hunter: Considering his dualistic, self-centered, competitive, and generally trivial motivations for hunting, abetted by his addiction to gadgets, motors, and other manufactured shortcuts, and no matter how much time, money, and effort he may invest in winning the "hunting challenge," I don't consider this type an authentic hunter. A devoted dabbler and dilettante, perhaps, but not a devoted hunter. Stephen Kellert agrees. Without the elements of challenge and competition, he predicts, the dominionistic/sport hunter's interest is unlikely to be sustained.

Of course, challenge and competition are not a priori bad. Kellert notes: "Seeking and stalking a large mammal can develop strength, prowess, cleverness, and technique. Together, the body and mind generate an enhanced physical and mental capacity." Indeed, as previously outlined, protracted participation in the intellect-building "upward-spiraling dance of predator and prey" (Shepard) deserves perhaps most of the credit for making us human. And too, personal challenge, in varying degree and according to various interpretations, is among the prime motivations of all hunters. And challenge implies competition.

Yet for most meat hunters and all "nature hunters" (soon to be profiled), any conscious sense of competition is focused more on skill and less on kill. It's only when the competitive drive expands to the level of killing to feed a sickly ego—as promoted, codified, and glorified by trophy hunting "clubs"—that it becomes morally reprehensible. As poet/hunter Jim Harrison points out: "Any spirit of competition in hunting or fishing dishonors the prey. It means that you are either unaware of, or have no feeling toward, your fellow creatures."

AMERICA'S OLDEST and most prestigious hunting trophy record-keeping organization is the Boone and Crockett Club (B&C), founded in 1887 by future U.S. President Theodore "Teddy" Roosevelt. Teddy's cofounder was famed naturalist and *Forest & Stream* editor George Bird Grinnell, backed by a select group of one hundred influential hunter/conservationists of the time (Stuyvesant, Pierrepont, et al.).

In those early, earnest days, B&C, working and networking through its well-connected membership, was arguably the single strongest force in halting the rampant decimation and extirpation of wild species through unregulated market and subsistence hunting. Happily, as U.S. Fish & Wildlife Service historian John Organ notes, "when Roosevelt became president of the United States, the Boone & Crockett Club [wildlife conservation] creed soon became national policy."

To cite one important example of its early good work, Boone & Crockett members, led by Grinnell, were responsible for persuading the federal government to enact the Yellowstone Park Protection Act of 1894—sponsored in Congress by B&C member John Lacey—which dispatched army troops to the world's first national park to control the voracious poaching activities ongoing there. By so doing, B&C was significantly, if not single-handedly, responsible for rescuing the American bison, the wapiti, and other then-endangered species from looming oblivion.

But the conservation activities of turn-of-the-century hunters were far broader than that. By the close of the 1800s, many native species were already extinct: Who among us is unaware of the blitzkrieg obliteration of the passenger pigeon, the apogee of American extinctions? Meanwhile, many other grand species—white-tailed deer, wild turkey, wood duck, pronghorn—were all going fast.

Today it's believed that North America has more deer than at any time since Christoforo Colombo washed ashore, to be followed close at heel by "Killer Cortez" and his pious ilk. And the wapiti, the second-largest and most regal of North America's five deer species, is now thriving throughout the western American wilds in states and provinces where, just fifty years ago, next to none could be found. In North America today, the wild wapiti population is estimated at more than a million animals, with more than a quarter of that lot enlivening my home state, Colorado.

Elk and deer are but two prominent examples of miraculous wildlife comebacks among dozens. And hunters and hunter/conservation groups have been, and continue to be, the leaders in promoting and financing this miracle. As former USDA Forest Service Chief Jack Ward Thomas once remarked: "If you want to do a species a favor, get it on the hunted list." If it's fair to say—as I believe it is—that the ultimate "right" we owe to wild animals is the right to freedom from human-caused extinction, then hunters and hunter's organizations are the preeminent "animal rights" crusaders of all time.

A second trophy hunting organization, the Pope and Young Club, is smaller, younger (founded in 1961), but modeled closely upon B&C—right down to the mathematical intricacies of "scoring" trophy heads for the record books. (Scoring formulas for antlered species comprise a convoluted combination of measurements, including main antler beam lengths and circumferences, number and length of tines, maximum spread between main beams, symmetry, and more.) But while Boone and Crockett accepts entries killed with any legal weapon, P&Y is reserved for bowhunters. Because there are far fewer bowhunters—and killing animals with bow and arrow is far more difficult than with a rifle (because you have to get so much closer, which involves a whole universe of intricate woodcraft and shooting skills the long-shot gunner need never master)—Pope & Young provides an ego handicap service by offering much lower score requirements than does B&C. Thus a "world record" for P&Y often scores far lower than the B&C world record for the same species.

Like Boone and Crockett, Pope and Young promotes "fair chase" hunting. That term, coined by Roosevelt and Grinnell, established rules of self-regulation among hunters who desire to enter trophies in the B&C books. Among the common-sense ethical basics of "fair chase" are proscriptions against shooting animals from airplanes or any motorized land or water vehicle . . . against the use of airplanes, boats, or land vehicles to herd animals toward shooters . . . against the use of artificial lights or night-vision optics . . . and against killing fenced-in animals or those otherwise rendered helpless. Ironically, and one of the fattest flies in B&C's "fair chase" pie, is that this latter proscription—against killing helpless animals—doesn't apply to bears and cougars that have been treed by hounds, thus trapped, immobilized, and effectively rendered helpless, except in cases where the hounds are wearing electronic radio-tracking collars. Moreover, the current drift of club politics portends further weakening of the already minimal "fair chase" standards for both these leading headhunter organizations.

But why wait? For those eager trophy "collectors" unwilling to abide by even the foundation-level ethics of B&C and P&Y, there's always Safari Club

International (SCI). Despite its much-touted "global wildlife conservation agenda," I see a whole lot bad in SCI and damned little good. In fact, Safari Club International and its members have the onerous distinction of embodying most of the attitudes that antihunters love to hate. When I hear someone attempt to defend SCI, I know I'm hearing a fool.

For telling examples of SCI's shortcomings, I'll defer to hunter, conservationist, wildlife ethicist, and renowned investigative journalist Ted Williams. For years Williams has tracked SCI's multifaceted sins, reporting them in his *Audubon* "Incite" column and elsewhere. In the following excerpts from an open letter, Williams summarizes a representative few of the atrocities recently committed by SCI's leadership and members: "Safari Club International . . . is a tireless opponent of the Endangered Species Act [ESA]. . . . At the 1996 Outdoor Writers [Association of America] Conference in Duluth, I sat with several hundred of my fellow scribes while then-president of SCI, John Jackson, droned on about how awful was the ESA and how it inconvenienced him and his fellow members in their quest to acquire the biggest horns. It was excruciatingly embarrassing for every genuine conservationist in the room."

Next, quoting himself from a 1992 issue of *Audubon,* Williams considers the hunting ethics of a former president of SCI's "Ark-La-Tex" (Arkansas/Louisiana/Texas) chapter:

> Dr. Sonny Milstead, an orthopedic surgeon from Shreveport, Louisiana, . . . made national TV news this past September when the Fund for Animals [a powerful and active animal rights/antihunting group] obtained a video tape of his canned cat hunt at a game ranch near Fredericksburg, Texas. Dr. Sonny approaches the trophy as it reclines trustingly on the ground. Dr. Sonny is protected by backup gunners. When Dr. Sonny shoots, the lion leaps to his feet, clawing dirt. You can see a divot fly from his flank as the heavy bullet slams home. Dr. Sonny fires twice more; the lion dies. Nothing in the least illegal about that. [At least not in Texas, exotic game-ranch capital of North America.] But the tape also shows Dr. Sonny shooting a penned Bengal tiger as it rests beneath a tree. At the first shot the big

cat gets up and runs to the right, dragging its shattered hind quarters. When Dr. Sonny shoots again it somersaults three or four times. Dr. Sonny cautiously approaches, prods the dead trophy with his gun barrel, then flashes the thumbs-up sign.

By way of an update, Ted Williams adds this hopeful postscript to that hopeless event: "Unfortunately for Dr. Sonny, Bengal tigers are an endangered species, and shooting even a tame one is illegal in this country. According to court records . . . he was ordered to pay a fine of $20,000, stripped of his hunting privileges inside and outside the U.S., and sentenced to five hundred hours of community service."

What else? Williams points out that "the *SCI Record Book* contains page after page after page of trophies killed inside game-proof fences." And "of seventeen entries in the category of introduced North American wild boar," he continues, "all seventeen are from a game ranch in Nova Scotia," where so-called hunting is conducted "in an enclosure that averages seventy-five acres, and the 'wild boars' are fed commercial hog grower all winter."

Even among the loathsome likes of SCI members, it's only the worst of the dominionistic/sport subset—you'll *never* see this among meat or nature hunters—who literally buy their "trophies," executing penned or fence-enclosed wildlife, including imported "exotics." Often these pathetic "shooter" animals are selectively bred to produce large, ornate antlers (or horns, or skulls, or tusks), have pet names ("Goliath" and the like), and are tame enough to eat from your hand. Weighed on any valid moral scale, such fat-cat "canned hunters," along with the "game-ranch" operators who pander to them, are lower than a meat poacher's hemorrhoids. Poachers, at least, have to expend some effort, demonstrate some skill, assume some risk, and generally kill for food, not ego.

———

BY SELF-DEFINITION, any "hunt" that requires no hunting is no hunt at all. Yet the privatization of wildlife for profit—euphemized as "alternative livestock ranching" (or, in the vernacular of some regions, "game farming")—is a boomer growth industry. Moreover, it's eagerly endorsed by state departments of agriculture throughout Canada and the American West (with the noble exception of Wyoming)—even as it's decried on pragmatic as well as moral grounds by ethical hunters, including a majority of trophy hunters, and professional wildlife managers.

In addition to providing canned hunts, common game-ranching money-makers include the breeding and sale of pedigreed animals, the hacksaw "harvesting" of blood-engorged, velvet-stage antlers for the lucrative Asian aphrodisiac market, meat sales (primarily to "exclusive" restaurants), and even "futures" (the prenatal sale of the unborn offspring of pedigreed parents).

Dangers of game ranching include the spread of such captive-bred killers as bovine tuberculosis and chronic wasting ("mad cow") disease to wild populations, genetic pollution of wild populations by escaped exotics, wildlife habitat lost to game-proof fences or overtaken by exotics, and a variety of criminal activities among game ranchers—all well documented.

While some eighty-five percent of Americans approve of meat hunting, according to Stephen Kellert, only fifteen percent endorse trophy hunting. Frankly, I'm surprised the latter approval is even that high. And surely almost no one—save those who do it themselves or profit from it—would publicly endorse canned killing. So why is it legal? All the usual reasons: Landowner rights. States rights. Institutionalized Cartesian dualism. Money. Politics. The American Way.

"What stories would they tell?" wonders *Mule Deer* magazine editor Scott Stouder regarding today's pay-to-play wildlife killers. "From the campfires of the ancients to our living rooms, stories have been the glue that binds generations of hunters. . . . It was through the power of story that I first heard [expressed] the love of mountains and animals. And it was by that power that I eventually came to love them myself. What love would come from the stories told of shooting animals within fences?"

Love of anything beyond self is a foreign concept, I suspect, to this pathetic perversion of humanity. As the legendary Canadian wildlife biologist/philosopher C. H. D. Clarke notes: "It is, of course, self-evident that there must be pervert hunters, and even fishermen, just as there are pervert clergymen, or boilermakers. No group is exempt, in spite of Xenophon, and we have to watch out for the pervert who deliberately takes up hunting. Primitive man knew and feared such persons, as likely to bring disaster on the tribe by offending the spirits of the animals hunted. When I was north [in Canada], there was still living an Eskimo who had been blinded deliberately by his people for just such a reason."

Not all canned killers are perverts, of course; many are merely stupid.

BY COMPARISON to the likes of Safari Club International, therefore, Boone & Crockett and Pope & Young are saintly. Yet its proud heritage of wildlife conservation and fair chase ethics notwithstanding, even B&C, through the keenly competitive, scorekeeping, award-bestowing mentality it glorifies, is promoting a plethora of ego-driven sins of stupidity. Moreover, its attempts to limit those sins through the turn-of-the-century doctrine of fair chase, measured against today's world of rapidly evolving public ethics and runaway technological gadgetry, are increasingly antiquated and woefully inadequate.

What to do? What if trophy hunting groups, via some unforeseen miracle, acquiesced to mounting public pressure and deleted hunters' names from their record books? What then? Many of the most unsporting "sports," as Stephen Kellert suggests, would probably abandon trophy hunting for other pursuits more congenial to their staggering egos. But if that were to happen—if the major trophy hunting clubs ever put an end to scorekeeping—they would soon likely cease to be . . . only to be replaced by new groups with new books and even lower ethical standards. And they know it.

And, too, to lessen the current trophy fever among "sport" hunters would be to lessen the trophy-quality profits currently being harvested by the hunting industry and media. And *they* know it. Consequently, industry and media eagerly support trophy hunting and record keeping and will fight to keep both.

One especially ludicrous trickle-down of all this is that trophy clubs, industry, and media, working wallet-to-wallet, have spawned a smarmy blight of huckster hunting heroes. Watch for these smug, smiling faces on cable TV's *The Outdoor Channel,* in amateurish hunting and fishing videos, in "sporting" magazines—and, of course, in outdoor clothing and equipment catalogs, especially such biggies as Cabela's and Bass Pro Shops— eagerly promoting their own egos as well as the products and services of their sponsors. "Average" hunters, these sponsors hope, will emulate their heroes by purchasing garages full of celebrity-endorsed products and popping small fortunes for the exotic guided hunts these globe-trotting goofballs generally get gratis.

Another spin-off of commercially orchestrated outdoor hero worship is that the average hunter is made to feel embarrassed if he fails to kill or to bring home anything less than a "book-quality" trophy. It follows with predictable economic logic that industry and media also are cheerleaders for the chant that more and more hunters must be recruited and indoctrinated with Chicken Little antihunter paranoia in order to "protect what's right." To the contrary—as *Bugle* conservation editor Dave Stalling was among the first to point out—in order to assure the future of hunting, we don't need more hunters; we need *better* hunters.

By way of concluding this sordid sidetrip, outdoor writer Ted Kerasote, in his apologia *Heart of Home,* sums up the image problems that Stephen Kellert's dominionistic/sport hunters create for honorable hunting, along with that clueless sub-subculture's umbilical connection to Mother Culture:

> Unfortunately, it has been dominionistic/sport hunters, even though they represent less than 40 percent of America's hunters, who have often set the image for the rest of the hunting community. Despite hunters' best efforts at educating the public about the hunter's role in conserving habitat and species, it is this group's behavior that the public remembers when they hear the word "hunting." Not only are this group's actions highly visible, but as a group they may very well represent more of America's hunters than Kellert's study leads us to believe. Indeed, they may represent a great many nonhunters. The developer who fills a wetland, homeowners who spread toxic herbicides on their lawns, every one of us who supports monoculture forests, agribusiness, and factory farms participates in a type of dominionistic mastery over wildlife and nature. . . . On the other hand, the dominionistic hunter's actions are visible, premeditated, and often discomforting; they are, however, in keeping with the fundamental beliefs of the culture that has bred him.

Standing in proud contrast to both the coldly utilitarian meat "harvester" and the clueless, egoistic "sport hunter" is Kellert's third category: the eighteen percent minority he calls naturalistic/nature hunters. Years before reading Kellert's report, I had already identified this singular subset, which I dubbed *spiritual* hunters. Other fitting titles include real, authentic, and true hunters—all indicating a level of involvement far surpassing mere legality or even "fair chase."

Kellert's data showed the typical nature hunter to be younger, with higher education and income, than the other two groups. This category also included more women. Unlike sport and meat hunters, nature hunters were interested not only in hunting but in such diverse nature-based activities as camping, backpacking, even bird-watching. (In the first of several ironic conjunctions of opposites, this same set of outdoor interests, minus hunting of course, was shared by many of the most ardent antihunters in Kellert's study.) Nature hunters also hunted more often than members of the other two hunter groups—"perhaps suggesting," reasons Kellert, "a stronger commitment . . . to the activity."

In summarizing the characteristics defining nature hunters, Kellert notes that they had far and away the highest "knowledge-of-animals scale scores" among all those tested—hunters, nonhunters, and antihunters alike. Further: "The desire for an active, participatory role in nature was perhaps the most significant aspect of the nature hunters' naturalistic attitude. They sought an intense involvement with wild animals in their natural habitats. Participation as a predator was valued for the opportunities it provided to regard oneself as an integral part of nature. The experience of the hunt was appreciated for its forcing of awareness of natural phenomena organized into a coherent, goal-directed framework." In contrast to meat and sport hunters, nature hunters regarded their prey as "an object of strong affection, respect, and, at times, even reverence. In pursuing the prey, not only were its habits and abilities learned, but a vicarious sense was achieved of how it experienced its environment."

At this juncture, Kellert quotes from José Ortega y Gasset's *Meditations on Hunting*. The wise old Spaniard observes: "When one is hunting, the air has another, more exquisite feel as it glides over the skin or enters the lungs; the rocks acquire a more expressive physiognomy, and the vegetation becomes loaded with meaning. All this is due to the fact that the hunter, while he advances or waits crouching, feels tied through the earth to the animal he pursues."

Elsewhere in his insightful *Meditations,* prefiguring both the names and weaknesses of two of Kellert's three hunter types, Ortega cautions that "hunting cannot be defined by its transient purposes—utilitarian or sporting. They remain outside of it, beyond it, and they presuppose it."

When I use the term "hunter" positively, it's Stephen Kellert's nature hunter I'm referring to—the same person I interchangeably think of as a spiritual hunter, a true hunter, a *real hunter*. As friend and adviser Florence Shepard recently observed: "If fewer than twenty percent of hunters today truly know and value wildlife and nature, then hunting's in trouble."

And indeed it is. Yet even as a minority, we're talking about some three million men and—increasingly—women who view and conduct their hunting as something of a sacred pilgrimage.

As Vance Bourjaily notes in *The Unnatural Enemy:* "There is in [nature hunting] not sport so much as self-renewal; an acknowledgment that one is of natural origin and belongs first to a world built by forces, not by hands. [Spiritual] hunting gives shape and purpose to this sojourn, becoming a rite of simplification."

Embodying Bourjaily's, Kellert's, Ortega's, and my own ideas of the nature/spiritual hunter is my Oregon e-mail friend, Mitch Caldwell, who wrote me last fall:

My elk season here ended weeks ago, and I'm just now coming out of my annual post-season depression. It's always a difficult time for me, after elk season ends, and it has nothing to do with whether I brought game down or not. Rather, I miss involving myself with the woods and the wildlife in a way that's far more meaningful and gratifying than is possible when I'm out there just to see what I can see. The "edge" is gone, and I passionately miss it every autumn.

Even so, my hunts this year were good; great and pure as always. And as usual, I didn't bring home any of the wapiti I so love and love to hunt. Nor did I see nearly as many elk as I usually do, and was bummed when I went to one of my favorite spots on the east side of the Cascades, in the Umatilla National Forest, and found it cow-burned almost beyond recognition. No wonder the elk were scarce. I don't think the forest managers had any idea how many cattle were in there; they do now!

The elk this year presented no signs of rutting behavior the entire September archery season, which is typical here, if nowhere else. Instead, they went crazy the first two weeks of October, after I'd gone home. The last time I heard a bugle was back in 1994, which was also the last time I took an elk—a beautiful, hefty bull whose antlers had six points on one side, five on the other—my most prized natural work of art.

That 1994 bull, in fact, was the only elk I've killed in fifteen years of hunting. Most years, I have opportunities, if I want, to shoot several young bulls, and cows by the dozens. This year, however, I had only a few cows, and no bulls, amble by within spitting distance; quite disappointing not to get a chance to spend more time up-close with my wild friends. —Yours, Mitch

Motivated by such pure feelings, and with just one bull elk in fifteen years of hard hunting, Mitch Caldwell knows more about the true meaning and joys of hunting than all the sad fools listed in all the trophy record books extant.

————

A FINAL QUOTE from Stephen Kellert recaps his triad of hunter types and transitions us into an examination of their philosophical counterparts:

> The nature hunter . . . felt the need to confront and ratio-
> nalize the death of the animal. Motivated by genuine . . .
> respect for wildlife, the nature hunter faced the paradox of
> inflicting violence on a world that was the object of great
> affection. In contrast, [meat] and [dominionistic] hunters
> did not evidence this affection . . . and the kill was more
> easily justified by [utilitarian] goals. . . . The nature hunter,
> in contrast, professed a deep affection for wildlife, and this
> paradox [between feelings expressed by nature, dominion-
> istic, and meat hunters] was one aspect of the antihunter's
> skepticism of any hunter's claims to possess great feeling for
> animals, even the ones they killed.

Of course, the foundation of Kellert's research is now nearly a quarter-century old. What about today? Kellert and I share a mutual conviction that the utilitarian/meat hunter category has lost its numerical majority, due to increased urbanization, education, and income nationwide. As this decline is no doubt compensated by an increase in the other two hunter types, the question becomes: Which group is growing faster—sport hunters or nature hunters?

Searching for answers to this significant query, I found an article in a 1997 issue of *North American Hunter,* written by Mark Damian Duda, head of the noted research and polling organization Responsive Management, which specializes in outdoor recreational topics. According to Duda: "In a nationwide telephone survey of hunters, we found that 43 percent . . . hunt primarily for the sport and recreation, 25 percent hunt primarily for the meat, 21 percent hunt to be close to nature, and 12 percent hunt to be with friends and family." Since Duda's categories vary from Kellert's, it's impossible to draw statistical parallels. Even so, at least a slight increase in the number of nature-hunters is suggested.

But what about all the others—the majority who will never be spiritual hunters? As early as 1958, C. H. D. Clarke had recognized this paradox. And Clarke's solution, I believe, is just as valid today . . . and still awaiting implementation: "It is no good trying to make one of the basic activities of the human race . . . the exclusive property of a small cult. We have to let them hunt, even though we thereby include those who debase sport. They are the few. Of the rest, we must agree that few have the knowledge or perception to fit into nature as a hunter should. They are, however, willing and eager to learn. . . . By helping them we help ourselves, the game, and the whole world of nature." Helping both hunters and antihunters learn to "fit into nature," as all humans should—this is my hope and my motivation.

———

A LITTLE FARTHER along this bumpy trail of words, I'll address, from an intimately personal perspective, what Stephen Kellert calls the nature hunter's love/kill paradox. But first, it's appropriate to see what Kellert's research reveals about that apparently growing segment of the American public whose values lead them to damn even the most honorable hunting as at least "paradoxical" and at worst "psychopathological": the antihunters.

5

Antihunters: A Scientific Profile
(with Personal Conclusions)

> Hunters are on this continent a large company, and we are
> not likely to be deprived of our sport in a hurry; but if one
> reasonable voice speaks against us, we owe it to ourselves to
> seek the truth.
>
> —C. H. D. Clarke

IT WOULD SEEM—statistically at least—that a majority of the American public is quite "reasonable," even charitable, in its perception of hunting. According to a nationwide survey conducted by Responsive Management (RM) in 1995, "73% of Americans approved of legal hunting (40% strongly approved while 33% moderately approved), while 22% disapproved." Further, RM found that "most Americans believe hunting should remain legal. Eighty-one percent of Americans agreed that hunting should continue to be legal. Fifty-three percent strongly agreed, 28% moderately agreed, 3% neither agreed nor disagreed, 6% moderately disagreed, and 10% strongly disagreed."

Among the six percent who moderately disapprove of the continuation of hunting, those I've spoken with disapprove not so much of hunting per se but of specific behavior and techniques—baiting and hounding are often cited—that they find unethical and obnoxious. The ten percent who strongly disagree with hunting's continuation, however, are philosophically opposed to all hunting. And in America a ten percent minority—assuming it's well organized, well funded, and highly motivated, which the animal rights movement is—can squeak loudly enough to attract serious political wheel-greasing. To reject their concerns outright, as extreme "hunters' rights" groups are wont to do, is personally dishonest and politically ingenuous. In the words of Starker Leopold: "We cannot overlook the wave of antihunting sentiment or charge it to the rantings of an impotent, lunatic minority. It has, in fact, become a real political force, not only in the state and in the nation, but in international wildlife affairs."

As a passionate hunter who strives, as Dr. Clarke advises, to "seek the truth" as a solid base for my opinions and actions, my stance on "the anti problem" is this: By granting serious attention to *reasonable* criticisms of hunting, reasonable hunters can help to disenfranchise their unreasonable critics. But where are those reasonable hunters? And where is that voice of pragmatic accommodation? Here on the sharp tip of a new millennium—which many fear may be the *last* millennium for democratic hunting (or democratic anything, for that matter)—a majority of hunters continue to reject, or at least ignore, any criticism of their "sport" as profoundly unreasonable. At the same time, too many blind-sided hunters thoughtlessly brand all those who offer criticism, constructive as well as deconstructive, as "the enemy." Further, should a *hunter* criticize any aspect of hunting, that hunter is not only the enemy but a lowly "turncoat," "traitor," and "dissembler" to boot.

In all instances of true-believerism, whether pro or anti, as Aldo Leopold noted decades ago, "the wish is too obviously father to the thought. They [true believers] represent merely the age-old insistence of the human mind to fix on some visible scapegoat the responsibility for invisible phenomena which they cannot or do not wish to understand."

In the case of extreme hunters, that scapegoat is criticism (helpful as well as hateful). In the instance of animal rights extremists, feminist scholar Mary Zeiss Stange suggests that adherence to veganism is a "*sine qua non* for radical ecofeminism." It follows, says Stange, that hunting provides for this same zealous minority "a root metaphor for patriarchal man's rape of nature. Indeed, from [the radical ecofeminist/vegan] vantage point, hunting and rape are virtually interchangeable." At both poles, as Arthur Conan Doyle's "Sherlock Holmes" advises, "it is a capital mistake to theorize before one has data. Insensibly one begins to twist facts to suit theories, instead of theories to suit facts."

Which brings us back to C. H. D. Clarke, who observes that "what we have to regret most often is the human perversity that regards truth as slander where its acceptance means changing a previously advocated practice. . . . Clearly, the moral and ethical requirement is for fact-finding and then fact-facing." In pursuit of that bifurcate goal—and given the increasingly inflammable ambiance of the hunt/no-hunt controversy—it is essential to recognize the clear-cut distinctions between intractable dogmatic fanatics, as opposed to constructive iconoclasts, *on both* sides of the battle.

To help us recognize these essential distinctions—and the truths such recognition can reveal—I offer the following scientific/academic overview of antihunter types and motivations, comprising Stephen Kellert's work, my personal conclusions, and a little help from some friends. (Please pay good attention, as a quiz will follow.)

———

CONSCIOUSLY or subconsciously, people have always classified animals as "good" or "bad," generally meaning useful or threatening, and treated them accordingly. And always, the distinction has been circumstantial, subjective, and culturally driven. If you were a Paleolithic hunter/gatherer, a "good" animal would be one you'd like to eat. A "bad" animal would be one who'd like to eat you. In the modern agricultural worldview, much the same is true. Except that now most good (tasting) animals have been domesticated, while most "bad" animals—those that might eat your chickens or corn, thus threatening your livelihood—have been exterminated, remain free but feared and hated, or, as in the cases of skunks, raccoons, and porcupines, are at least resented.

So it has always gone. And so it goes today—as embodied in the American livestock industry's vicious, illogical, and ecologically destructive war on coyotes, cougars, bears, and other "vermin." In this war of ancient and abiding enmities, ranchers and farmers are eagerly aided—not only by tax-funded professional "predator control" agents in the employ of federal and state agriculture departments—but by a minority of deluded fun-killers calling themselves "varmint hunters." But that's another and far darker topic deserving a book of its own.

The public likewise distinguishes between good and bad animals. The former usually include pets, livestock, and wild herbivores; wild carnivores and predatory omnivores, including two-legged hunters, tend to comprise the latter.

Since the hunting controversy lives largely at the extremes of prohunting and antihunting sentiment, both edges are inflexible: problems and solutions are viewed not as matters of degree, not as negotiable, but as absolutes; all or nothing. The transparent truth is that neither side is substantially correct, while neither is utterly wrong. But try telling that to one who proudly displays a bumper sticker proclaiming *"I Kill Hunters for Fun and Sport!"* (marketed, ironically, by a California firm calling itself EnvironGentle)—or, representing the other extreme, *"Buy a Gun. Piss Off a Liberal!"*—and see how far you get.

In the opening pages of his report, Stephen Kellert quotes from a doctoral dissertation whose author, William Shaw, thinks that hunting's most enthusiastic attackers and defenders both tend to "assume that contrary opinions are due to shallow thinking or naivete. . . . Conflicting views [in fact] are often based on legitimate philosophical differences."

More bluntly: Militant antihunters view all hunters as knuckle-dragging, beer-swilling, blood-lusting, nature-raping cretins. This view is expressed vividly by Joy Williams—among the shrillest, most aggressive, and self-assured of animal rights voices—who proclaimed in *Esquire:* "For hunters, hunting is fun. Recreation is play. Hunting is recreation. Hunters kill for play, for entertainment. They kill for the thrill of it, to make an animal 'theirs.' . . . The animal becomes the property of the hunter by its death. Alive, the beast belongs only to itself. This is unacceptable to the hunter."

Tragically, this extreme perception is substantially correct in a minority of instances. But to Williams and those who share her blanket views, it's universal. Meanwhile, militant "hunter's rights" zealots brand anyone who criticizes any aspect of hunting as an evil "anti"—and generalize all antis as urban, mostly female, hyperemotional, and woefully uninformed about the workings of nature and scientific wildlife management.

And in a minority of cases, *this* "extreme" perception also is substantially correct—as evidenced by the Williams quote. Similarly, self-proclaimed radical ecofeminist Linda Vance advises those women whose lives and feelings and politics she would direct that "passionate convictions, beliefs from the heart, can always get us through the hard times when reason and argument fail."

No matter which dogma is barking, the parallels are obvious.

———

ACCORDING TO STEPHEN KELLERT, in the late 1990s hunting "constitutes the pastime" of between fifteen and twenty million Americans. Of that total, some ninety percent were men; an equivalent male majority filled hunting and fishing organizations. Not surprisingly, statistics in these areas vary widely if not wildly: hunters are prone to report higher estimates, anti-hunters lower, both perhaps hoping to create a self-fulfilling prophecy. Kellert's figures at least represent objectivity, if not pinpoint accuracy.

Another objective source is the public-opinion polling group Responsive Management, which reports a 1996 total of fourteen million licensed hunters, or seven percent of the U.S. population. "Within the U.S. popula-tion," RM says, "thirteen percent of males aged sixteen years and older hunt, and about one percent of females hunt." Moving forward a year, to 1997, the *Wall Street Journal* reports fifteen million hunting licenses sold.

No matter which figures we choose to use, we must add children, seniors and the disabled, Native Americans hunting on tribal lands, and others who, in many states, are not required to buy licenses—thus accounting for Kellert's twenty-million top-end estimate.

———

WOMEN COMPRISE about seventy percent of antihunters and (depending on your source) seventy to eighty percent of animal rights and animal welfare groups.

Rights versus welfare: there is a difference. Responsive Management dis-tinguishes the two in this fashion: "Animal welfare supports the humane treatment and responsible care of animals that ensures comfort and freedom from unnecessary pain and suffering"—adding that "most American support animal welfare." By contrast, says RM, "the animal rights movement dictates that the use of animals for any human benefit is wrong. . . . One should not fish or hunt, wear silk or wool, sweeten food with honey, eat eggs or dairy products, play sports with a ball made of leather, patronize zoos, aquariums, or pet shops, or watch movies or TV shows that employ trained animals. . . . Most Americans do not agree with the animal rights philosophy." And even among the fifteen percent of Americans who do hold an animal rights attitude, notes RM's Mark Duda, only three percent actually refuse to use animals for any purpose.

It follows that "animal rights," unlike animal welfare, can fairly be equated with entrenched antihunting sentiment. As one who's morally and spiritually offended by many aspects of culturally sanctioned cruelty to animals—industrial chicken, pig, veal, and egg factories; trivial and overtly commercial cosmetic and redundant medical experimentation on animals; mistreatment and neglect of pets; and genetic god-playing—I am, de facto, an animal welfarist.

Even so, as a nature hunter, nature lover, and neo-animist, I find it impossible to endorse any animal rights group, even though we may agree on many animal welfare issues, if that group universally decries hunting, good as well as bad, which all such groups inevitably do.

———

BEFORE GOING FURTHER, I should further clarify my own admittedly amorphous position regarding hunting and antihunting. In the recent past, antihunters have chosen to interpret my stance as rabidly prohunting, even as the "hunters' rights" crowd have branded me as "an anti in hunter's camouflage." Neither side has a clue. But both prove the point that nobody these days likes a fence-straddler.

And that's exactly what I am: a dedicated hunter who shares many sentiments and concerns with hunting's more reasonable and informed critics. In fact—if I didn't have a hunter's heart and the experience and insights gained through a lifetime of hunting and hands-on nature study—I could, I suppose, become an anti myself—given the mountain of cultural crap that currently accompanies modern so-called hunting. The line is just that thin, stretched tight along boundaries of personal values and philosophies, making it easy to understand why so many nonhunters, and many former hunters, are so easily swayed in the anti direction.

As Ortega notes, *Homo sapiens* evolved as a predatory omnivore equipped with both carnivore's fangs and vegetarian's molars, thus combining "the two extreme conditions of the mammal, and therefore [going] through life vacillating between being a sheep and being a tiger."

But lest we forget: While the tiger is a creation of natural selection, the sheep is a creation of human selection. I cast my lot with the tiger.

———

As STEPHEN KELLERT NOTED of hunters, it's equally "simplistic and even foolish" to speak of antihunters as a group. Yet, as he does with hunters, Kellert identifies a few striking "anti" generalities. To begin: "Antihunters were crudely identified by their strong agreement with the statement 'Hunting for sport is wrong.'"

Most of Kellert's antis, confirming the common perception among hunters, were in fact female and urban, residing in cities of a million or more, clustered along the Pacific and mid-Atlantic coasts. Antis also reported "significantly less" experience with animals in an agricultural setting than did others (hunters and nonhunters) in the sample.

Further, even though the welfare of animals in general, and pets in particular, ranked as vital concerns among Kellert's antis, these same people, in league with the dominionistic/sport hunters they despise, achieved "among the lowest knowledge-of-animals scores of any group included in the study." In yet another ironic parallel with sport hunters, "it appeared that antihunters manifested more fear and lack of interest in wildlife" than average Americans. Yet these folks purport to care more deeply for animals than the average American and claim to know animals' needs and wants better than others.

At the opposite extreme, in a philosophical confluence of odd bedfellows, both nature hunters and antihunters "appeared to perceive an equality and kinship, rather than a hierarchical-dominant relationship, existing between humans and animals."

———

KELLERT'S DATA led him to split antis into two basic types—humanistic and moralistic—based primarily on emotional and philosophical grounds. Contrasting the humanistic and moralistic philosophies, Kellert reports that they

> appeared to be quite different antihunting viewpoints and
> often did not occur in the same person. The humanistic
> objection to hunting was motivated largely by strong affec-
> tion for and emotional identification with animals and a

related concern for the presumed pain and suffering experienced by hunted animals. The moralistic antihunting posture, in contrast, stemmed more from broad ethical and philosophical notions regarding the proper behavior of human beings, which included the notion that killing for sport was inherently wrong and evil. In other words, the humanistic antihunting posture was typically quite animal-centered, whereas the moralistic antihunting viewpoint was far more related to the presumed impact of hunting on human social and ethical behavior.

Among Kellert's humanistic antihunters, pets were personified to the point of being essentially indistinguishable from humans. In turn, this "pets as people" mindset expanded to enfold all animals—especially such "charismatic megafauna" as deer and other big, beautiful mammals "with whom empathetic identification was easier to achieve." Such higher-order creatures were pictured by humanistic antihunters as being capable of experiencing "feelings of fear, terror, and pain similar to those that might characterize a human being's reactions [if] placed in a similar situation." Kellert goes on to note:

> Identifying with the presumed emotional and mental experience of individual animals represents a critical basis for a humanistic perspective of nature. This emotional identification figures in many people's perceptions of large mammals. Occasionally, difficulties and distortions arise from strong humanistic associations, such as when extreme affection for individual large mammals results in undue and dysfunctional anthropomorphism (e.g., the so-called "Bambi" syndrome). . . . Well-known children's stories like Winnie the Pooh, Goldilocks, Bambi, and Little Red Riding Hood, for example, have depicted bears, wolves, and deer in highly anthropomorphic ways intended to promote various morals and codes of conduct.

(Sweet little baby-faced Bambi. How can we ignore him in any serious discussion of hunting and antihunting? We cannot. But Bambi must wait his turn, a few chapters along, for he deserves more than a mere passing shot.)

Not only are humanistic antihunters intent on promoting "various morals and codes of conduct" to which Stephen Kellert refers. They are its most unquestioning and enthusiastic proselytizers. As one who hunts, kills, and eats wild animals, but also studies, loves, and works to protect wildlife and wild places, I was long perplexed by proclamations from extreme animal rights spokeswomen (in the two instances I'm recalling, both in fact were women) that they would rather see a magnificent species—in the first instance, African elephants; in the second, white-tailed deer—go extinct rather than have them continue to be "managed" through hunting. Now, finally, thanks to Stephen Kellert, I understand that such sentiments are restricted to a specific philosophical extremity even among antihunters. To wit: "The proposition that the well-being of a species was more important than the experience of the individual [animal] had little relevance to the humanistic antihunter."

THE MORALISTIC ANTIHUNTER'S HATRED of hunting and hunters, by contrast, was driven more by concern for humanity, less for the hunted animal. As Kellert summarizes this viewpoint: "An important tenet of the moralistic antihunting position was the notion of life as sacred and the infliction of death for any reason other than necessity and survival as degrading. . . . Many moralistic antihunters identified with Albert Schweitzer's concept of reverence for life. As one moralistic antihunting respondent [said]: 'Albert Schweitzer, he's one of my heroes.'"

He's a hero of mine as well. No red-blooded neo-animist such as I could fail to applaud such gilded Schweitzerisms as: "By respect for life we become religious in a way that is elementary, profound, and alive." And: "Man can no longer live for himself alone. We must realize that all life is valuable and that we are united to all life. From this knowledge comes our spiritual relationship with the universe."

Yet a study of the life and words of Albert Schweitzer reveals that he was starkly inconsistent and, at times, coldly cruel in the allocation of his avowed "reverence for life." When weighed against his own utilitarian/ agricultural background (he was a happy swineherd as a boy), as well as his views, his conduct, and his carnivory, the good doctor was not only disingenuous in his pious condemnations of hunting, but downright hypocritical.

AT HIS MEDICAL MISSION at Lambaréné, in the province of Gabon, in what was then French Equatorial Africa, Schweitzer kept domestic animals for food and wild animals as pets. And in one infamous incident, a pet, when it displeased its kindly master, was promptly demoted to food. I'm referring to Josephine, a vaguely domesticated wild boar and Albert's special pet, who, along with sheep, chickens, and other barnyard beasts, had free run of the Lambaréné compound—at least until Josephine "intruded on Sunday services" at the mission church, earning the doctor's ire and prompting him to write to a superior "that because of all this, I should kill her."

When, soon after, Josephine ate some chickens, at Albert's direction she "was enticed into the hospital, tied up, and expeditiously and artistically slaughtered." Which leaves us to wonder how punitive butchery, albeit "artistically" executed, jibes with "reverence for life."

Even that flagship phrase, "reverence for life," is tinged with Schweitzerian hypocrisy. These three words (in the original German: *Ehrfurcht für dem Leben:* Honor/Awe for Life) came to Albert one day, as he recalls, "when at sunset, we were making our way through a herd of hippopotamuses" and "the iron door yielded; the path in the thicket had become visible." Reverence for life, all life, was the key. Not long after, however, the moody doctor mused about killing a local hippo should it threaten his garden. Thus we see reflected in Albert Schweitzer the agrarian Everyman's coldly utilitarian distinction between "good" and "bad" animals.

And so he continued, our good doctor, personifying the Arabian proverb "A mouth that prays, a hand that kills." Let's consider a few more instances. In accord with his proclamation that an ethical person "is careful not to crush any insect as he walks," Albert made much ado of not stepping on pismires. But the larger and more troublesome "traveler" (army) ants, being "serious enemies," he destroyed wholesale. And in that destruction, disregarding his own counsel to "not use insecticides for killing the poor creatures," he exterminated army ants en masse with Lysol. "Thousands of corpses lie in the puddles," he boasted.

Regarding "recreational" hunting, Schweitzer asks impatiently: "When will we reach the point that hunting, the pleasure of killing animals for sport, will be regarded as a mental aberration? We must reach the point that killing for sport will be felt as a disgrace to our civilization."

Certainly, most of the sport hunters Schweitzer encountered in that time and place—ignorant executioners of lions and leopards and elephants and hippos for nothing but "trophy" body parts—*were* in fact disgraceful. Yet Albert erred in painting all hunters and all sorts of hunting with this same dirty brush . . . except, of course, for himself and his own small hunting pleasures. Once, spying the "alarming" and "hateful shadow" of a spider creeping down his wall, Schweitzer notes with grudging admiration that it was "much bigger than the most magnificent ones I had seen in Europe. After an exciting hunt, it was killed."

And there are more displays of mendacious irreverence for life. When a leopard entered the compound and killed some chickens, Schweitzer demonstrated not only a European peasant's fearful superstition regarding predators—noting that the chickens' breasts had been "torn open," Albert pronounces that "only the leopard kills in this way, for his first desire is to get blood to drink"—but also, as with Josephine, a flair for creative revenge: "One of [the dead chickens] was filled with strychnine and left lying before the door. Two hours later the leopard came back and devoured it. While the leopard was writhing in convulsions it was shot." In condoning such "necessary" cruelty, even to such blood-lusting demons as leopards, we can rest assured the benevolent healer was, as always, "regretful."

Nor is this the end of it. To Schweitzer, gorillas were "real monsters" about whom one biographer reports the doctor "retailed misinformation." Snakes were "enemies" toward whose extermination Albert kept a gun ready at hand (handy also for pot-shooting various birds of prey who might steal a beloved chicken). "Pet" gazelles and other captive animals were kept in small pens and led about the compound on chains—which is reminiscent of one of Schweitzer's most unlikely contemporary bedfellows: fellow hunter-hater Herr Heinrich Himmler.

———

BUT ENOUGH PETTY PICKING on that poor dead semisaint, Dr. Albert Schweitzer. He was a better man than most. My point is merely that no one can escape personal responsibility for the deaths that sustain our lives, not even a semisaint. To believe otherwise is your choice. But to press those beliefs on others is moral hypocrisy, even if you're Albert Schweitzer.

Paul Shepard captures the essence of Schweitzer's distinctly human strengths and weaknesses when he writes:

No one doubts the inspiring leadership of [Schweitzer's] mission. His compassion extends to all the natural world, and he recognizes that our intellectual and philosophical substantiation of this feeling is weak. His remedy is to extend human justice into primeval patterns of animal life. Because this justice is ecologically naive and foreign, its effect is to destroy or alter those natural communities. This alteration, in the long view, may be detrimental to humanity. Perhaps no European has more convincingly identified and criticized the homocentricity of Western philosophy. But the difficulty is not merely one of omission. The remedy does not lie in turning the machinery of ethics as they are now understood upon the animal world. A successful solution will have to include understanding of the web of animal life and respect for its own native community laws. . . . Reverence for Life does not provide a working pattern that conserves the natural environment of which particular kinds of living things are part.

Reinforcing Shepard's views in this is Ontario biologist and hunting ethicist C. H. D. Clarke—the "Canadian Aldo Leopold." In his classic essay "Autumn Thoughts of a Hunter," Clarke sharply chides Schweitzerian "reverence for life":

Any concept of life that does not comprehend the whole organic cycle is inadequate. The reluctance to accept death, evidently a predominant Schweitzer characteristic, reveals an unseeing devotion to the vital spark. It is death that makes it glow, measure for measure. . . . Schweitzer is right in affirming that the decay of civilization is caused by the lack of a proper relationship between man and other organic life, and its restoration can come only when such a relationship is established. . . . Man has lived, and in some places still lives, in harmony with nature, and the hunter and angler still cling to strong lines that connect us with the harmonious past.

Well, some of us do. Yet I fear that Clarke and Shepard's academically sophisticated, deep-ecological, conservation-biological, neo-animistic musings aren't likely to convert many devout antihunters, humanistic or moralistic, who, like their extreme opposites in the "rights" camp, crave a more literal, simplistic, predictable, personally compliant, and controllable world.

————

By way of giving name, voice, and personality to Kellert's moralistic antihunter, I can cite personal example. I hope that in doing so I'm not motivated by a desire for Schweitzerian revenge (though as a "psychopathologically impulsive" hunter, it's hard to say for sure). In any event: In the fall of 1998 my book *Elkheart: A Personal Tribute to Wapiti and Their World* was published. As usual for "nature books," it was greeted by resounding critical silence. One review, however, and a strongly favorable one at that, did appear in *High Country News,* the American West's best-known environmental journal. While the reviewer was careful to point out that *Elkheart* was not a "hunting book" but a narrative-driven natural history and personal appreciation of elk—exactly as its title and subtitle announce—he nonetheless chose to focus on the hunting-related aspects of the book. These bits included a vigorous defense of natural predators and ethical human hunters as natural predators, an outspoken condemnation of morally outrageous hunter conduct, a rant against commercial elk "ranching," and a heartfelt explication of my personal views on hunting, both pro and con.

Predictably, given the contentiousness of hunting, the review sparked a lively exchange in the letters columns of subsequent issues of *High Country News.* Opening the barrage were two vigorous ad hominem attacks on my semisaintly self written by men who had not bothered to read my book before offering their impassioned responses to it—notwithstanding that one of them omnisciently damns *Elkheart* as a "textbook example" of human evil.

But to the point: My first critic, Stephen Gies, exemplifies Kellert's finding that "the moralistic antihunter views killing for sport as the essential ethical difference between the hunter and meat-eating nonhunter." Mr. Gies, a supermarket carnivore, chastised me for killing my own meat rather than buying it shrink-wrapped. My reaction was merely a sad shake of the head and a recollection of *Time* writer Lance Morrow's query: "Who has clean hands? Surely not the consumers of the 38 million cows and calves, the 92 million hogs, the 4 million sheep and 7 billion chickens killed last year."

Thinking along the same lines as Morrow, a philosophical friend once wondered aloud: "What happens to a culture such as ours that avoids the direct experience of the kill? Psychologists would say that denial of such a fundamental guilt—that another life has to end in order for one's own to continue—sets the stage for self-righteousness. On a more conscious level, without the deep knowledge of human capacity for error that is part of the traditional hunter's experience, it becomes much too easy to think that we can *know* enough to perform right action."

Similarly, in *Woman the Hunter,* Mary Zeiss Stange observes that "the 'ethical vegetarian' who persists in ignoring the consequences of large-scale agriculture and the meat-eater who would rather not think about how a steer becomes a Big Mac, are in this regard equally self-deluded." Likewise, Paul Shepard prefers to shake his finger rather than his head at such hubristic illogic as Mr. Gies's. Shepard observes that "the interdependence of life . . . is likely to be obscure to those who turn the killing of food animals over to specialists who practice in secret. Those who fear death become politically and socially conservative and less tolerant of other species, other creeds, and any deviation from their own mode of life."

More creative and more provocative than Gies's mild missive was a subsequent missile fired at the journal's letters editor by Marc Gaede, of Pasadena, California, marking its author as—to borrow a phrase from Gaede himself—a "textbook example" of the moralistic antihunter. To these antihunters, as Kellert describes them, "hunting seems to be a form of violent antisocial conduct; the hunter is portrayed as sadistic; the activity is deemed likely to foster psychopathic tendencies. From this high moral plane, antihunting sentiment stems less from sympathy for the animal than from a fervent belief in the activity's intrinsically degenerative impact on people and society."

It was Mr. Gaede—a self-described anthropologist with a special interest in primatology and evolutionary psychology—who pronounced the unread *Elkheart* a "textbook example of displaced primate male-to-male aggression for females." Gaede goes on to profess that both humans and chimps inherited from a common ancestor a tendency to violence. With unblinking certitude and careful detail he lists the exact biological and social events that transpired among our prehuman ancestors and near relatives fifteen million years ago to bring this about. In sum, paraphrasing Mr. Gaede: As a result of male competition for estrous females, war evolved among our ancient ancestors. And when no war was on, "processes of ritualized war become all the rage," carried forward and exemplified today by the many and various ball sports. In addition to war and football, a third avenue for venting male sexual frustration/aggression also evolved—and that avenue is hunting. To heck, says Gaede, with such excuses for killing animals as procuring food: "Science can demonstrate that [male hunters] are, in fact, subconsciously killing other male humans because of competition for females." (Where, we're led to wonder, does that leave the ten percent of today's hunters who are women?)

The professor concludes his stern lecture with a flourish of male-to-male aggression of his own. Asserting that I am uncomfortable with my lowly social position because it restricts my "access to desired hierarchical females," he suggests that I consult with my "close female companions" on this matter.

I'll refrain from responding personally to Marc Gaede's darkly post-Freudian take on human evolution and gender relations. Nor will I embarrass my critic with the reactions of my "close female companions," some of whom confound the issue by being ardent hunters themselves. (Feminist scholar Mary Stange points out: "To the extent that hunting has served patriarchy and feminism as a root metaphor for men's activity in the world, Woman the Hunter is necessarily a disruptive figure; she upsets the equilibrium of the conventional interpretations on both sides.")

Instead, once again, I'll call upon Paul Shepard, who has good and wise things to say on so many important topics, always adeptly walking the fine line of passion unclouded by emotion—a line that neither I nor Marc Gaede can seem to tread. From *Coming Home to the Pleistocene:*

> The analogy of the hunt to warfare and crime is deeply wrong. . . . The modern heaping of abuse on hunters who "kill things" has escalated in the last years of the twentieth century. This rising hysteria about killing reminds us that the problem with death was never so intense among primal peoples—who participated in the great round—as it is

among those societies who dread it as a final calamity and strive to deny it and for whom it becomes a neurotic obsession. Psychological research indicates that moral codes are more rigid among people who dread death and whose inflexibility is projected into all kinds of social conservatism. When morality is premised on the escape from death, it is aimed at all those "causing" or participating in it. Most death in nature is invisible and, moreover, is accompanied by the fantasy that animals who are not killed (by people) go on living. The killing of one animal by another, so seldom seen, can be ignored or turned back into the unconscious. This repressed notion about nonhumans is released as fury against human hunters. . . . The decade of the 1980s witnessed a spate of essays on the "morality" of hunting. Focus on the ethics of hunting decontextualizes the subject. Its rhetoric of killing as evil, and compassion as its opposite, is abstract. Animal rights ethicists disembowel the subject the way a small mammal is collected for taxidermy. Having taken away the guts that connect the animal to its surroundings, there remains a shell, deprived not only of its own life but of the putrefaction that reintegrates the dead with the living.

C. H. D. Clarke adds that his own experiences as an explicator and defender of authentic hunting have "brought me references to what I now feel justified in calling, however unkindly, the *humaniacal* literature. It is easy to see the element of fanaticism. The true fanatic is very much convinced of his own righteousness, and sure that only the perversity of human nature prevents the whole world from agreeing with him. He also resents not so much the supposed sin itself as the pleasure that it gives."

But enough. To conclude on a note of balance, I'll give the last word to Duke University professor (and thoughtful antihunter) Matt Cartmill. Commenting on "The Bambi Syndrome" in *Natural History* magazine, Cartmill points out that "each side in this confrontation sees the other as a congregation of lunatics. Pro-animal activists regard hunters as macho gun nuts eager to prove their manhood by drilling anything that moves. For their part, hunters view their opponents as sentimental fluffheads, who dream of the natural world as a happy Eden and idealize wild animals as little innocent people in bunny suits."

Certainly we see ample examples of such lunacy at both extremes of this slippery subject. Yet, unlike the popular media, who thrive on stereotypes and feed on controversy, we shall not let the exceptions define the rules.

———

IN THE END, Ed Abbey was right (again) when he observed that people often come to love ideas more than they love life itself—in this case, *wildlife*. Even as the antihunting and hunters' rights congregations (especially the former) spend tens of millions annually to attack and counterattack in a no-win word-war of irreconcilable worldviews, wildlands and wildlife continue to fall victim to economic growth and so-called progress. Thus are both sides—each blindly devoted to its non sequitur cause—doomed to lose the real war by default, dragging the objects of their ostensible concern down with them. Award-winning nature writer and cultural anthropologist Richard Nelson lays it out cleanly when he warns that "after we've lost a natural place, it's gone for *everyone*—hikers, campers, boaters, bicyclists, animal watchers, fishers, hunters, and wildlife—a complete and absolutely democratic tragedy of emptiness. For this reason, it's vital that we overcome our differences, find common ground in our shared love for the natural world, and work together to defend the wild."

I agree passionately. And so does a growing coalition of "fence-straddling" others. In fact, if all these pages can be reduced to a single message, Nelson's is it. Yet regarding combative true believers, pro and anti, I'm not holding my breath. You just can't reason with unreasonable people.

———

AND NOW, in conclusion, we come to the quiz I threatened earlier. I'm going to present extracts from the official hunting position statements of a representative sampling of animal *rights* groups (which, you'll recall, are staunchly antihunting)—mixed with the positions of a sampling of animal *welfare* groups, whose memberships, in varying degrees, comprise functional coalitions of hunters, nonhunters, and antihunters. Your task is to determine which is what: Which groups are primarily *ideological*—clearly humanistic and/or moralistic—in their purported concerns for animals and theirs views on hunting? And which are, by comparison, *practical?* Bottom line: Which persuasion, in your opinion—welfare or rights—is more valid and useful and why?

(The following excerpts are lifted from the National Shooting Sports Foundation brochure, *The Hunter in Conservation.* Budget figures, where shown, are from *Animal People* magazine, December 1998, as reported in the Wildlife Legislative Fund of America's *WLFA Update,* February 2000.)

- "The *American Humane Association* [AHA] is opposed to the hunting of any living creature for fun, a trophy, or for simple sport. The AHA believes that sport hunting is a form of exploitation of animals for the entertainment of the hunter, and is contrary to the values of compassion and respect for all life that inform American Humane's mission. . . . [AHA] considers sport hunting a violation of the inherent integrity of animals and disruptive of the national [sic] balance of the environment through human manipulation, and calls for positive action to prevent such cruelties."

- "*Defenders of Wildlife* is neither an antihunting nor a prohunting organization, but most of its 80,000 members are nonhunters and their concern is with the restoration and protection of all species of wildlife and their habitats."

- *Friends of Animals, Inc.:* "Hunting is cruel. It is deceitful. It is socially unjustifiable. It is ecologically disruptive. Friends of Animals opposes hunting in all its forms." [1998 budget: $4,514,292]

- "The *Fund for Animals* is unalterably opposed to the recreational killing of wildlife. There is no reason to kill and cripple animals in the name of fun. Our society has progressed beyond the days of the gladiatorial games when people enjoyed inflicting pain on others. . . . Hunters—a mere 7 percent of the public—have 100 percent control of our wildlife. The Fund for Animals seeks equal representation for the majority of Americans who oppose the killing of animals for sport." [1998 budget: $5,445,455]

- "The *Humane Society of the United States* [HSUS] is strongly opposed to the hunting of any living creature for fun, trophy, or for sport, because of the trauma, suffering, and death to the animals which results. The HSUS also opposes such killing because of the negative effect upon the young who may learn to accept and live with needless suffering and killing. The

HSUS believes that a civilized society should not condone the killing of any sentient creature as sport." [1998 budget: $36,633,759]

- "The *National Audubon Society*, since its origin at the turn of the century, has never been opposed to the hunting of game species if that hunting is done ethically and in accordance with laws and regulations designed to prevent depletion of the wildlife resource. . . . [Yet] we do not advocate hunting. This is no contradiction, though some people seem to think it is. Our objective is wildlife and environmental conservation, not the promotion of hunting. We think lots of the justifications for hunting are weak ones, and too often exaggerated for commercial reasons, and we do not hesitate to say so when the occasion calls for it. But this does not make us antihunting. We are pushing people to think more clearly about these problems."

- "The *National Wildlife Federation* [NWF] is composed of millions of members [over four million, in fact, making NWF the largest private conservation organization in the world] and associates whose primary interest is in the conservation of our nation's renewable resources. Although our members and affiliates have many and varied opinions of how these resources might be best utilized, hunters and nonhunters alike support our broad conservation objectives. We support hunting because, under professional regulation, wildlife populations are a renewable natural resource that can safely sustain taking. Although we understand the moral philosophy of those who feel that hunting is wrong and that wildlife populations should be protected completely, the real and fundamental problem facing wildlife is not hunting but . . . habitat degradation and destruction. [NWF] therefore, is committed to conserving wildlife habitat. To accomplish this objective, hunters and nonhunters should unite efforts to preserve wildlife habitat, the key to wildlife variety and abundance."

- *Sierra Club:* "Wildlife and native plant management should emphasize maintenance and restoration of healthy, viable native plant and animal populations, their habitats, and ecological processes. Acceptable management approaches include both regulated periodic hunting and fishing when based on sufficient scientifically valid biological data and when consistent with all other management purposes and when necessary, total protection of particular species or populations. Because national parks are set aside for the preservation of natural landscapes and wildlife, the Sierra Club is opposed to sport hunting in national parks." [Complete statement]

- "The *Wilderness Society* recognizes hunting as a legitimate use in wilderness areas, national forests, and certain wildlife areas, subject to appropriate regulation for species protection." [Complete statement]
- *World Wildlife Fund* (WWF): "As with most charities of its kind, there are widely divergent views on the issue of hunting among the supporters, board, and staff of World Wildlife Fund; however, the organization itself takes no position, either pro or con, on hunting. [WWF] recognizes that responsibly conducted hunting can be an appropriate wildlife management tool. . . . On the other hand, given the irrevocable consequences, [WWF] opposes hunting which might adversely affect the survival of threatened or endangered species. It is unfortunate that adversarial relations between organizations over the issue of hunting have been to the detriment of a broader common interest in maintenance of the biological heritage of the globe."

The World Wildlife Fund's take on hunting and antihunting—and its unifying call for a hunter/antihunter truce on behalf of the beautiful wildness both sides proclaim to love—echo my take and my call closely.

Yet pragmatism cuts only so deep. To comprehend the beating bloody heart of the matter, we must push boldly ahead, into the realm—both numinous and luminous—of *spirit.*

6

Hunting for Spirituality:
An Oxymoron?

[*Spiritual* and *sacred* are terms that refer] to those inexplicable relationships and processes that govern existence.
There is no reason sacredness cannot be manifest in *any* circumstances whatever, or in all circumstances, even if some
are more numinous than others.

—Paul Shepard

As ONE WHO MAKES an earnest (if not always successful) effort to think (if
not always speak) objectively, I've never faced a more emotionally con-
tentious challenge than substantiating and articulating the spirituality inherent to true hunting. So troubling is this topic that few, on either side, feel
comfortable even talking about it.

One problem is that many people today, men in particular, are uneasy
with the lexicon of "secular" spirituality and feel reluctant to discuss or even
think about spiritual issues outside the culturally codified bounds of a litur-
gical setting. Moreover, spirituality is broadly viewed by males as bound in
a Gordian knot with emotionalism. And to reveal emotion, many men
fear, is a sign of weakness. Also, the recent rash of New Age mysticism has
tainted secular spirituality and language with a sour taste that many, myself

included, interpret as muddled metaphysical insipidae. For these reasons and more, spirituality is not a topic you often hear discussed around hunters' campfires.

Nor, for reasons less sympathetic, will you often see such words as "sacred" and "spiritual" in the pages of today's commercial hook-and-bullet magazines, where the term "big bucks" is a corporate double entendre. Nor is the fractured "sporting community" alone in its unease with talk of spirituality in hunting. To evangelical animal rightists, any use of the S-words in association with "blood sport" is likely to elicit outraged howls of "Oxymoron!"

And frankly, I see their point. After all, how can killing *ever* be considered a sacred act—an experience that makes life more meaningful and fulfilling and helps to shape better human beings? Especially when it's conducted under such seemingly trivial titles as "sport" and "recreation" and presented, as it so often is in the outdoor media, as a technological war on wildlife: *If you're a deer, you're dead!*

How can spirituality be aligned with such as this?

It cannot.

True spirituality, as experienced by the nature hunter, not only has naught to do with such a body-count mentality, it refutes it.

As in most failures of communication between differing views, part of the problem lies in what *Bugle* editor Dan Crockett calls "the shaky architecture of nomenclature."

Over a century ago, the term "sport" was adopted by thoughtful "sportsmen" to distinguish self-regulated "recreational" hunting from ecologically untenable subsistence and market hunting, both of which were rampant at the time and amoral, at best, in practice. Their ends—food on the table or money in the pocket—justified in the hunters' minds any means.

Thus, in the beginning, the term "sport" and its various permutations, as applied to recreational hunting, were strongly positive, as espoused by Ortega and others, signifying self-imposed rules of restraint and decorum afield, guided by a strident conservation ethic. As recently as midcentury and beyond, "sport" as a positive modifier for hunting continued to be endorsed by such meditative conservation pioneers as Aldo Leopold, who recognized that "there is value in any experience that exercises those ethical restraints collectively called 'sportsmanship.' Our tools for the pursuit of wildlife improve faster than we do, and sportsmanship is a voluntary limitation in the use of these armaments. It is aimed to augment the role of skill and shrink the role of gadgets in the pursuit of wild things."

Even today, many intelligent and ethical hunters—such as writers Jim Harrison and Thomas McGuane, a pair of prominent examples—embrace "sport" and "sporting" as suggestive of absolutely honorable conduct in hunting and fishing. Both Paul Shepard and Stephen Kellert occasionally use the terms "sporting" and "sportsmen" in strongly positive context. Among the "sporting" media, the richest example of nominal loyalty to the intrinsic honor of "sport" is *Gray's Sporting Journal,* one of the most honest (and easily the most artful) of commercial hunting and fishing magazines.

I too, until quite recently, have matter-of-factly enlisted "sport," even "blood sport," as a positive or at least neutral modifier for hunting—if for no better reasons than tradition and convenience. Even in these pages I'm often obliged to quote the term in positive context from the writings of others, which I do with full understanding of "where they were coming from." Yet I'm trying to kick the habit. I've finally acquiesced to the fact that today this particular S-word is inextricably entangled in the mass-media-trained mass mind with commerce, competition, aggression, divertimento, beer, and balls.

While the term "recreation" as applied to hunting lacks the increasingly negative cultural connotations of "sport"—and in fact is an accurate if incomplete modifier in most cases—it too is restrictive and often misleading. Like "sport," the word "recreation" trivializes what should be—and, for those who do it right, is—far more than mere entertainment. As Mark Duda and his cohorts at Responsive Management note in *Wildlife and the American Mind:*

> Hunting, however much a recreational activity, can only be
> fully comprehended if it is understood as a complex cul-
> tural phenomenon closely linked to naturalistic values,
> one's identity, and the American family. To the nonhunter,
> hunting is often viewed as just another recreational activity.
> But research clearly indicates that there is something much
> deeper. Hunting is an important social and psychological
> activity for hunters. Hunting is a powerful and meaningful

pursuit, seemingly above and beyond other forms of recreational activities. . . . Hunting is not just recreation.

I'm hardly alone in my struggles with the shaky architecture of hunting nomenclature. For instance, the U.S. Fish & Wildlife Service's John F. Organ, et al., suggest that the most appropriate and least offensive modifier for modern hunting is "regulated." While this may in some ways be an improvement over "sport" and "recreational," it's too bureaucratic for my tastes. Since market hunting is dead and subsistence hunting is rare outside bush Alaska and Canada, why not just call it . . . hunting? Which, in turn, requires us to define "hunting" and its kin.

THOUGH EXTREMISTS at both edges, pro and anti, will eagerly argue otherwise, attempting to reduce a massively complex issue to a simplistic right or wrong, one fact is clear: Not all hunting is equal. Not all approaches to hunting and not all hunters are good. Not all approaches to hunting and not all hunters are bad—although to an observant public the latter must often seem the case. Anyone who pays attention can quickly amass a long list of rude, stupid, and manifestly shameful examples of hunter behavior, most of which fall into the category of care-less-ness: disregard for the safety and sensibilities of others, disrespect for the hunted animal, disrespect for the land (littering, driving off-road, leaving gates open, fouling surface water), public displays of ignorance and arrogance, and, of course, more.

Meanwhile, other examples of hunter misconduct—involving dubious products and services and, always, profit—are eagerly endorsed by a voracious outdoor industry and its venal allies among the commercial hook-and-bullet media. For examples aplenty, just browse through any commercial hunting publication and you'll see dozens, scores, hundreds of products advertised and editorially hyped, many of which clearly offer unethical advantage to unethical hunters. In business, hunting and otherwise, legal means moral. Consequently—to the hunting industry and media flacks and the hunters who support them—the more that's legal, the better. To hell with ethics and self-respect. This blatant commercial corruption of hunting, of course, links hunting ethics to social ethics: hunting as cultural metaphor.

And this is another important point that's often overlooked by non-hunters as well as antihunters: Hunters are not a species apart. Hunters are people—fathers and mothers, daughters and sons, saints and sinners alike. Which is to say: If a person is a slob hunter, he's predictably a slob in every other regard: work, family, community, traffic, even what passes in his life for spirituality. There's nothing inherent to the act of hunting that promotes moral erosion or incites blood-lust, as hunting's harshest critics, having no personal experience, choose to believe. Rather than creating personalities and worldviews, hunting merely reflects them, good and bad, as shaped by the overarching human environment.

In order to reform hunting, therefore, we must reform media, marketplace, and, ultimately, Mother culture. But that's a topic far too immense and depressing for the here and now.

———

IN A SHAMELESS ATTEMPT to put a human face on Stephen Kellert's statistical entity the naturalistic/nature hunter and make him real and personal, I now offer my meager self as a more or less typical—some will say extreme—specimen. I know many more. But I can speak best from personal experience.

I've been "outdoorsy" and semiferal since I could walk; I was, you could say, "born to be wild." Consequently, across the mounting years, I've enjoyed many kinds of hunting, pursued in many arenas, insofar as all honest hunting offers the blessings of fresh air, varied exercise, earth rather than concrete under your boots, sky rather than walls embracing your vision, opportunity to practice such ancient skills as tracking, calling, and interpreting the secret habits and habitats of wildlife—plus, of course, the possibility of bringing home a sacramental portion of delicious (if carefully handled and prepared) and nutritious wild meat, the food that made us human.

Yet not all hunting and hunting habitats hold equal allure to nature hunters. When I'm hunting in notably human-altered environs—with the peripheral presence of roads and fences and livestock and buildings and power poles and noise—I'm unavoidably distracted and annoyed. Others I know feel the same. To attain the personal epiphanies, aesthetic bliss, visceral emotions, and introspection that inform true hunting—nature hunting, spiritual hunting—I need, first and foremost, the solitude, silence, and beauty of wildly natural surroundings. Toward this end, I seek my happy hunting grounds far from cacophonous "progress": places where the scenery and tranquillity alone are ample justification to be there, and where I'm unlikely to encounter other humans. (For all the same reasons, I can't separate my passion for hunting from my passions for human population control and wilderness preservation.)

Further, the nature hunter needs a credible challenge: to work hard, to suffer and sacrifice at least a little, and to fail more often than succeed. By approaching hunting in this adventurous spirit, when I do bring home some meat I know I've earned it, both morally and physically. It follows that I'm an equipment and technology minimalist.

STEPHEN KELLERT was not yet born when the pioneering conservationist Aldo Leopold first recognized and chastised the fathers of Kellert's dominionistic/sport set. After praising the skills, self-reliance, effort, humility, and naturalistic outlook inherent to traditional, authentic hunting, Leopold went on to damn the postwar outbreak of "gadgeteering" among hunters:

> And then came the gadgeteer, otherwise known as the sporting-goods dealer. He has draped the American outdoorsman with an infinity of contraptions, all offered as aids to self-reliance, hardihood, woodcraft, or marksmanship, but too often functioning as substitutes for them. Gadgets fill the pockets, they dangle from neck and belt. The overflow fills the auto-trunk, and also the trailer. . . . I have the impression the American sportsman is puzzled; he doesn't understand what is happening to him. . . . It has not dawned on him that the outdoor recreations are essentially primitive; atavistic; that their value is a contrast-value. . . . The sportsman has no leaders to tell him what is wrong. The sporting press no longer represents sport; it has turned billboard for the gadgeteer.

A decade later, by way of seconding Leopold's distress with the billboarding of the North American "sporting press," C. H. D. Clarke noted with disgust that "we can envy the Europeans the intelligent level of their sporting journals! Ours started well but now approach the level of the so-called 'comic book.'. . . The emphasis, we can add, is on what are called 'gadgets' and 'gimmicks.' The ruling hand is not nature's God, but Mammon. The trend is towards moronic mediocrity."

This creeping disease of the human soul—which Leopold first diagnosed within the American hunting media in the 1940s, whose cancerous growth Clarke denounced in the 1950s, and whose insidious erosion of traditional hunting values Stephen Kellert so carefully documented in the 1970s—metastasized during the 1990s into the autocannibalistic technomalignancy that's devouring the heart and soul of hunting, and America, today.

Playing Moloch to the sporting press's Mammon, the Internet as well has now turned "billboard for the gadgeteer." For an outlandish example consider the Fall 1998 issue of *www.HuntingNet* magazine ("the official publication of the world's largest hunting website"). Announcing the issue's theme is a bold gold headline writ large across the chest of a young, Ramboesque cover model: "Takin' the High-Tech Road." Gadgets fill the pockets of this virtual bowhunter's computer-designed camouflage clothing and, indeed, "dangle from neck and belt." Hammering the hard-sell home, each goodie is highlighted in a close-up, captioned photo: GPS (Global Positioning System, an electronic device allowing the dilettante outdoorsman to wander anywhere and back again without basic map and compass skills), mechanical broadhead (an arrowhead whose cutting blades collapse and remain closed for greater speed and accuracy in flight, popping open on impact), night-vision binoculars (use your imagination), electronic rangefinder (optically measures the exact distance from hunter to target), etcetera.

Further, taking full advantage of most states' laughably liberal definitions of "primitive weapons" (including bows, muzzleloading rifles, and in some states crossbows)—users of which are granted longer and generally more favorable seasons—our synthetic nimrod is equipped with a space-age compound bow constructed almost entirely of high-tech synthetics and elaborately configured with cams, cables, pulleys, overdraw, sight, stabilizer, and other modern "primitive hunting" gadgets that make drawing and holding the bow at "full cock" less strenuous, the shooting more accurate, and requiring, withal, far less skill and practice as an archer.

Finally, in his "free" hand our high-tech hero carries a portable self-climbing tree stand—an essential aid for outfoxing those clever cornfield Bambis and Falines, in that it allows the hunter to sit in relative comfort high in a tree, above the deer's normal line of sight and scent.

But for all of this, we're left to wonder, where is our gadgeteer's Game Finder Pro—a "hand-held infrared detector [that] senses temperature differences as small as 1°F. . . . Advanced computer technology for incredible range and accuracy. Range: up to 1000 yards in open terrain, up to 200 yards in wooded areas." (Yours for $269.99 in the Redhead Fall 1999 Hunting Specialists Catalog.)

And where is our man's Bionic Ear—an "omni-directional microphone [which] provides a broad range of pointed sound pickup." And where is his matching Bionic Ear Booster—a "parabolic 12-inch reflector [which] works with Bionic Ear. Pinpoints sounds and increases amplification"? ($79.99 and $39.99, respectively, from the same Redhead catalog.)

To save time that might otherwise be wasted studying animal vocalizations and learning to imitate them with your own voice (or with a traditional, breath-powered, hand-held call), surely our virtual hunter needs a Johnny Stewart long-range electronic game caller, featuring "a rechargeable battery which provides 20 to 25 actual hours of calling time. Comes complete with everything you need, including speaker and 25-foot speaker extension cord, and recharger." (This device is hyped in Cabela's Fall 1999 catalog for $184.99.) For an additional $8.99 each, you can become the proud owner of calling tapes for mallard, goose, wild turkey, wild pig, fawn, and many more. And with the optional auto cigarette lighter power cord ($12.99), you don't even have to open the pickup door or get up off your ass to "hunt."

And why fool around actually *hunting* for deer? For only $399.99 you can own a Buckshot 35 Infrared Game Scouting Camera. As the Redhead catalog proclaims: "Many hunter-hours are spent on the stand every season waiting for the right deer to come by. This camera not only reveals if there are bucks in the area, but it also leaves no doubt as to the quality! Buckshot 35 gives you a glossy color photo with a time or date imprint, so you'll know when he's on the move. . . . The Buckshot 35 is designed to simply take pictures, day or night, of anything warm-blooded that moves in front of it."

And to make darn sure that something warm-blooded will in fact be around, exactly when and where you want it to be, an absolute essential is a Moultree automatic game feeder. "Make the game hunt you! These feeders let you modify or actually create new wildlife feeding patterns so you can attract and hold concentrations of deer, turkey, and other game." Available from Redhead for as little as $49.99 for the Monster Hanging Feeder (holds fifty pounds of feed with automatic timer), up to $299.99 for the Premiere Magnum (holds two hundred and twenty pounds of feed with programmable timer that automatically dispenses bait as often as four times per day). And for the hunter on the move, there are ATV and tailgate-mounted feeders. *Now* we're hunting, Texas-style! Other near-mandatory gadgets available from any number of "hunter supply" outlets scattered all up and down the *www.HuntingNet* "High-Tech Road" include compact two-way radios (allowing two or more hunters to gang up on game), mechanical string releases (replacing obsolete fingers for a more precise, triggerlike bowstring release), scent-proof clothing (keeping body odor from escaping), synthetic rattling antlers (imitating two bucks sparring—to attract rutting deer), and *so much more!*

Surely no self-respecting "primitive-weapons" hunter would venture afield without such techno-tools as these. Is our cover model perhaps underequipped? More likely, all that unseen booty is stashed in his ATV, waiting on its tilt-bed trailer behind his SUV, over on the far-side of the suburban cornfield.

The net effect and unabashed intent of all this *stuff*, of course, is to reduce the need to develop traditional woodcraft skills, exercise patience, or exert physical effort and to help assure consistent kills—in effect, to take the *hunt* out of hunting; to *buy* success rather than earn it.

Dazzled by the *www.HuntingNet* cover display of techno-wizardry, the would-be hunter feels compelled to open this "magazine"—wherein, promise the cover teasers, he'll be educated and entertained with such woodsy campfire treatises as "Using the Internet to become a better hunter," "How high-tech made the difference," and "In review: High-tech tools."

As Ontario hunter, hunting ethicist, and wildlife biologist Michael Buss points out, the modern gadget industry strives to "insulate outdoorsmen from the very elements which should spice their outdoor experiences. To view some outdoorsmen, one would think they were headed for an outing on the moon."

———

NOR ARE GADGETMANIA and marketplace Mammon even the worst of it. The worst—the absolute worst—is a clueless killer-cult that pays thousands to execute captive, increasingly biogenetically engineered "trophy" animals and dares to call it hunting.

These are the bottom-feeders of the testosterone-drenched subset Stephen Kellert calls dominionistic/sport hunters. These drooling, incompetent few—a tiny minority even among the headhunting crowd—literally buy their "trophies of a lifetime," paying big bucks to indulge in the shooting-gallery killing of big bucks in fenced enclosures. And increasingly, the victims in this bloody for-profit business—deer, elk, and other "game" species—are selectively bred for the specific morphological traits the bean-counters of trophy hunting so dearly value: horns, antlers, and skulls. Many of these ranch-raised captives have pet names and will eat from your hand.

———

THE PRIVATIZATION and genetic manipulation of wildlife for profit—euphemized as "alternative livestock ranching" (or, more Cartesian yet, "farming")—is booming throughout the rural United States and Canada. Moreover, it's eagerly endorsed by most state and provincial departments of agriculture. (One noble exception is Wyoming, where commercial trade in wildlife, in whole or in part, is staunchly prohibited.) Never mind that game farming and canned killing are roundly decried everywhere by ethical hunters, wildlife managers, and those relatively few among the nonhunting public who know of it and give a damn. It's good business, and business is America's business.

How can this be? Why is wildlife profiteering, including canned killing, so ubiquitously legal and so little protested in America today? Let us count the whys:

- "Landowners' rights," which proclaim that a man can do with his land as he damn-well pleases—and to hell with the neighbors, the ecology, the future.
- "States' rights," which allow local politicians—often back-slapping good old boys (and occasionally girls)—to ignore honor and logic and the national weal in favor of short-term profits for favored constituents.
- Culturally inculcated Cartesian dualism, which reassures us that the world was made for humans: since no animal has feelings or conscious awareness, we're free to abuse them as we please.
- Money.

- "Jobs." One of my favorite bumper stickers reads (somewhat ungrammatically): More jobs? What about less people?
- Politics.
- Business as usual.

And please remember this: Each time you enjoy a meal of venison, elk, or other "wild game" at a restaurant, you become an active participant in the obscene cruelty of game ranching. We already have enough domesticated species—known to geneticists as "goofies." Let wildlife *be* wild.

———

To DATE, the bio-engineering blight has yet to infect free-roaming wildlife or true hunting: no designer breeding of trophy-antlered elk or deer for "seeding" in the wild. So far, the bio-manipulation of wildlife remains limited to the private arena: hybridizing pen-raised "hunting preserve" pheasants to maximize their "sporting qualities"; transplanting embryos from one subspecies of elk into another in hopes of building a "super" subspecies; feeding nutritionally "hot" supplements to promote massive antler growth. ("Boone & Crockett in a Bucket," proclaims the headline of a full-page ad in *Bowhunter* magazine. "Want to start seeing some of those heavy-racked wallhangers on your property? Who doesn't? That's where new BioLogic Full Potential Mineral Supplement comes in.")

And, too, there's the sperm-banking of the semen of trophy males, coupled with artificial insemination of "factory mother" cervids. The stellar example here is one "30-30," among white-tailed deer the world's largest-antlered captive. (The terms "domestic" and "tame" are inappropriate.) In 1996, 30-30 was purchased "at stud" for $150,000. Today his semen is worth more than gold-plated cocaine. After being "electro-ejaculated," each load is divvied into several dozen test-tube doses that sell to breeders for $1,500 a squirt.

All of which brings us back to the cowardly minority who finance this Nazi Dr. Doolittle circus of wildlife horrors: people who eagerly "invest" $8,500 or more for the privilege of gunning down a designer-bred, custom-nourished "trophy" deer. Execution rights for a big bull elk can cost a whole lot more.

Delighted by all of this and dreaming of an ever more profitable future, profit-hungry wildlife geneticists—according to Mark T. Sullivan, writing in *Sports Afield*—"are working to map out the genetic code of whitetails so they can isolate the antler chromosome that will make genetic engineering for big-racked [deer likely in the] near future. With that map will come the possibility of cloned trophy deer."

And given that brave new bio-tech breakthrough—when it comes, which it will—why limit our God-playing to the private fenced pasture, where only the filthy rich can participate and profit? Already some "quality-conscious" trophy hunters are petitioning state and provincial wildlife agencies to adapt the wonders of bio-engineering to the genetic "enhancement" of public wildlife on public lands.

Should this ever happen, true hunting is dead. Naturally evolved wildness, likewise, will be no more. And where to then? "A world of made," warns e. e. cummings, "is not a world of born."

———

COLORADO BOWHUNTER Lane Eskew, who sent me the *www.HuntingNet* cover that so raises my dander, attached a note offering a nature hunter's (his own) take on the bio-tech bastardization of hunting—at the same time pointing us back toward where this all began: to the wisdom of Aldo Leopold.

"The Modern Bowhunter," Lane wrote in his note, referring to the Star Wars cover model, "sponsored by your local sporting goods store, sporting magazines, suppliers, advertisers, wildlife agencies, and most hunters. This makes me ill. It's not a joke!"

No, not a joke, Lane, a travesty—for the public image and self-respect of hunting, for the dignity of wildlife, and especially for all those who strive to be authentic hunters but are led astray by gaudy commercial hokum. As C. H. D. Clarke noted long ago: "Those who love hunting must . . . become eternally vigilant, especially as the most artificial aspects of human culture spread like a cancer."

This is not to suggest that nature hunters must be primitivists, or even strict traditionalists, renouncing all modern technology. Aldo Leopold was not and did not. And I, like other backcountry veterans, have learned through hard, occasionally dangerous experience the value of, say, modern synthetic "breathable" rain gear, which allows me to stay out in miserable weather in relative comfort and safety. Similarly, I carry a shirt-pocket binocular, not merely to search for game, but to enhance the overall nature experience: admiring both macro and micro scenery, spying on songbirds and other nongame wildlife, gawking at the golden-apple splendor of a fat full moon hung low in a blue-black crepuscular sky.

And so on. Including camouflage. I've worn camo, in fact, since before it was widely commercially available, much less stylish. And so did American Indians and other traditional hunters, as do serious wildlife photographers and field biologists, since "blending in" is essential to meaningful wildlife observation. And, too, it's darn hard to avoid: current choices in functional outdoor clothing are pretty much limited to camo at one extreme and flashy neons at the other. But recently, increasingly offended by the hyper-commercialization of the camouflage industry and increasingly annoyed by the paramilitaristic look that hunting, especially bowhunting, has lately taken on—I'm returning to somber, woodsy plaids, which seem to offend neither animals (so long as you're motionless, in shadow, and have a screen of foliage behind you) nor (of increasing import today) the watchful nonhunting public.

My premier outdoor passion is bowhunting for elk. For this demanding adventure, my "weapon" (the nomenclature is again shaky) is a traditional recurve (an ancient design whose tips curve forward, providing a bit more arrow cast, or thrust, than a straight-limbed longbow), artfully handcrafted from wood by master bowyer Patrick Ley, and bare-naked of any bells, gongs, or gadgets, and weighing only three pounds.

Such a basic tool—together, recurves and longbows compose the "traditional" archery category and are often referred to as "stickbows"—demands that I get breath-holding close to my secretive, super-alert quarry and practice the art of archery religiously in order to be reliably lethal, humane, and well fed. My reward for this extra effort is that getting close and getting good add greatly to hunting's challenges and gratifications.

In hunting as in life, I've come to realize that more is less. Whether hunting, fishing, hiking, camping, or canoeing, the more artifacts of technoculture we haul into the wilds, the less relaxing, challenging, focused, gratifying, adventurous, memorable, and, I believe, moral our time there ultimately will be—because the farther we are distancing ourselves from nature and our own innate wildness.

Wildness, solitude, simplicity: the Holy Trinity of nature hunting. Many authentic hunters do use technologically sophisticated tools—specifically, scope-sighted rifles and compound (cam-assisted) bows. Often this is because they can't arrange adequate practice time to become confident with more demanding weapons, and go sick at the thought of wounding an animal through incompetence. Perceived in this light—as moral necessity rather than immoral advantage—reasonable, nonelectronic hunting technology is difficult to decry. While hunting embodies a dual hierarchy of commitment and ethics, the two don't always run parallel.

———

THE BORDERS between Stephen Kellert's three hunter types, as noted earlier, are blurred and overlapping. For personal example: Like all nature hunters, I'm also a meat hunter, in that I kill nothing I won't eat and in fact select my prey largely for their flavor (along with local availability and other, less tangible qualities). This is basic. To kill and not to eat destroys all moral justification for hunting. Beyond being delicious, I prefer my prey big, beautiful, and unpredictable: creatures of the grand, charismatic sort you must study and observe for years in order to hunt well—and whose deaths you cannot so easily take for granted.

What else excites my personal passion for hunting? If an animal speaks a language I can to some extent decipher and vocally or mechanically (never electronically) imitate, all the better: the yelps, clucks, purrs, and putts of wild turkey hens (imitated with a mouth call made from a turkey wing bone, or with a wooden box call, where a "paddle" is scraped across a resonance chamber like a bow across a fiddle), the percussive gobbling of a wild turkey tom (this one I've mastered with a wild, head-shaking combination of mouth, throat, tongue, and cheeks), the eerie fluting of geese (wooden mouth call), the birdlike chirps of cow and calf elk, and (glory be!) the brassy bugling of September bulls. (In imitating elk I push my personal limits of ethical technology by employing breath-powered calls using "reeds" made of prophylactic rubber.)

Minus the neoprene, hunters have been impersonating their prey forever. And not only human hunters: I've read credible accounts of both European brown bears (grizzlies) and Siberian tigers consciously imitating the "roars" of red deer (Eurasian elk) in order to lure rut-crazed stags to ambush.

This is utterly natural.

———

IN THE END, everything turns on attitude and expectation. What you want from the hunt largely shapes *how* you hunt. What I want is an opportunity to play my naturally selected and (therefore) naturally moral role as an active participant in the most intimate workings of wild nature. I want to live the way humans were meant to live. And I want the meat.

"Above all," advises C. H. D. Clarke, "we should esteem in ourselves and in our companions that maturity of spirit which alone places the right to hunt beyond question." "Maturity of spirit." That's the ticket to moral indemnity: in the woods, in the office, at church, or at home. When I get out there and slowed down and tuned-in enough to perceive and appreciate even the subtlest elements of natural creation—the soft warm pastels of lichen on cold hard granite; a velvety fuzz of moss on a rotting log; the symphonic purling of a mountain stream; one red flicker feather—when I'm in that good old "savage" groove, I'm buzzing with the joy of life . . . and unafraid of death. Without the intercession of clergy, shaman, or psychotropic drug, I have stepped through the looking glass and into a primordially *sacred* realm. I have entered heaven on earth.

Indeed, any postmortem paradise lacking bugling September elk, fire-gold aspens rattling in a crisp autumn breeze, scarlet-throated trout leaping for joy in swift sparkling water, the wild hungry howls of wolves and coyotes, the preternatural wailing of loons on a moon-mirrored lake, the shotgun slap of a beaver's tail on a cloistered pond, the humbling *aliveness* that comes with the possibility of meeting a grizzly bear or mountain lion around the next bend along some shadowy forest trail, the gritty ecstasy of love on the rocks, the spicy bite of a hand-rolled cigar, the smoky bouquet of a cask-aged whisky—any heaven lacking such distinctly earthly blessings as these would be pure hell for me.

Yet why—I've often been asked—why can't thoughtful, nature-loving hunters attain this same level of neo-animistic bliss just by watching or perhaps photographing wildlife without looking to kill? It's a good question. And my instinctive answer is . . . instinct. To paraphrase English writer James Hamilton-Paterson, a camera gets in the way of the hunter's eye. Nor has Paul Shepard neglected to give this tough question due thought: Shepard concludes that "wildness cannot be captured on film; wildness is what I kill and eat because I, too, am wild."

Ontario biologist and hunting ethicist Michael Buss joins Shepard and Hamilton-Paterson when he reasons that true hunting "requires a concentrated searching and observing of the environment. The deliberate techniques make hunting a nature study which the pure naturalist can never emulate. To see the environment through the practiced eyes of a hunter is to identify not only the *characters* in nature, as a naturalist might, but also to experience the 'emotions' of those characters, without giving them unrealistic, human or Disneylike qualities."

Stephen Kellert affirms this subtle transformative merging process when he notes that for the nature hunter, in "pursuing the prey, not only were its habits and abilities learned, but a vicarious sense was achieved of how it experienced its environment."

Along this same line, Mary Zeiss Stange professes that "habits of eye are also habits of mind. The way one engages one's natural surroundings is also the way one engages ideas. The hunter's habit of mind has more to do with a mode of awareness, a discreet style of engagement, than with pulling a trigger or drawing a bowstring." Again we encounter Ortega's "alert man."

Moreover, nature hunters do rejoice in "just watching without looking to kill"—just as we always have. "Foraging peoples," Shepard tells us, "typically spend thousands of hours every year pondering and studying the animals around them." And authentic contemporary hunters are no different. I spend ten months of every twelve "just watching" wildlife, jubilantly "pondering and studying the animals around" me, without ever thinking of killing. Like the seasonally prescribed ritual it is, hunting has its bounds. And even when hunting, I sneak and peek and watch in wonder for weeks, while the killing, if and when it comes, kills only moments.

———

TRADITIONAL, LOW-TECH, one-on-one hunting fosters an intense and clearly primal level of alertness I'm rarely able to experience otherwise: what Ortega calls the "good hunter's almost mystical agitation." It's a vestigial sixth sense that's largely inaccessible to those culturally domesticated human animals who no longer participate in our species' traditionally active predator/prey relationship with wild nature.

An outstanding exception, however, the evolutionary flip side of the selfsame coin, is attainable by going solo and unarmed into grizzly country— becoming potential prey rather than dominant predator. Nothing, not even bowhunting, makes your senses and spirit quicker and more earth-connected. In grizzly country, the days are brighter, the nights are darker, the air is sweeter, and every moment is parturient with possibility.

While I frankly wouldn't care to feel like prey all the time—say, by living in a big city—an occasional dose of raw, edgy fear creates a primordial high that, in measured doses, can enhance both appreciation for life and acceptance of death. Tragically, the pristine quality of wildness required to support such magical creatures as grizzly bears is, like the great bruins themselves, on the brink of extinction.

———

ONE OF THE MOST INTENSE and memorable experiences of my life was a long black night spent canoe-camping with photographer Branson Reynolds in the Yellowstone backcountry in an area where grizzlies had congregated to feed on spawning native cutthroat trout. Bear sign was everywhere, and we had been officially cautioned, to the point of raw paranoia, about the "acute grizzly threat" in the area we were exploring. In fact, hiking trails had been closed and a horse-mounted trail-maintenance crew had fled the area. Consequently, neither Branson nor I slept a nod that night. Not a sound from the obsidian forest surrounding us, shroudlike, no matter how subtle, escaped our radar ears.

The outcome—no slavering monsters ever materialized—matters not a whit. It was our own personal Night of the Grizzlies, wherein myriad horror fantasies were conceived and dealt with. And for that humbling adventure, we remain thankful.

By contrast, here's a recent example, admittedly minor, of the subtle sort of connection I often experience as a direct result of maintaining a hunter's intimacy with wildness. This past spring, my old dog Otis and I were enjoying a long evening walk up the mountain above our shotgun shack. At a place where a big toothy boulder snags out from a high rocky rim, providing an exhilarating balcony from which to view the forested valley below, we stopped to rest and meditate.

Soon, as was my habit at the time, I fell to pondering loss. It was March 14, the tenth anniversary of the premature death of my friend and hero Edward Abbey . . . and just one week since my wife Caroline (and therefore I) had suffered the wholly unexpected theft, from stealthy cancer, of an old best friend from high school. Additionally, both Caroline and I had serious health concerns. So there I sat, bushwhacked and befuddled by a sinking sense of mortal angst . . . when suddenly, from across the quiet valley, rose the explosive spring serenade of a wild tom turkey: *Gobble-obble-obble!*

At that distance, it was not loud. Yet Otis sprang to his paws with visible animal excitement. And so did I. In that instant, not only were my self-pity and creeping despair wholly forgot, I caught myself smiling. And when you smile, you *always* feel better. It was the first gobble of spring, signifying for me a rite of seasonal passage—like the flowering of the first spring beauty, the ringing arrival of the first hummingbird: a promise, a benediction, . . . a reminder.

"Life goes on," I assured Otis. "No matter what."

"Life is *good*," I assured myself.

And with that we were up and on our way again, spiritually (and thus physically) reinvigorated with the joyful energy of hope.

While this moment may seem insignificant and even silly to some, to me it's anything but. And that's my point. Beyond the ranks of serious hunters, few would even have noticed that faint, far-off call of the wild. Fewer still would have sensed its spiritual significance. Why and how might this be so?

Believe me when I say that getting pumped for the approaching turkey season—a predatory urge to track down and kill that randy old gobbler—was not part of the moment. Yet had I not learned something of the intimate lives of *Meleagris gallopavo* through years of studying and hunting them; had I been unable to see in bright memory's eye the passionate reds, whites, and blues pulsing like a neon marquee across that carbuncled, prehistoric face . . . the iridescent copper sheen of his inflated feathery breast . . . the Indian-headdress splendor of his wide-fanned tail; had I been uninterested in imagining the priapic emotions running wild through that old bird's hormone-drunken being—without such hunting-bred intimacy, that impromptu cathartic confluence of wild-turkey nature and wild-human nature would never have come to bless me.

It's the nature hunter's "minding of the environment," says Shepard, "the fluid quality of his attention, and the habits of alertness and acuity [which] link him in participation with all of creation." And this sentiment reminds me of these poetic lines from Leopold's *Sand County Almanac:* "The deer hunter habitually watches the next bend; the duck hunter watches the sky-line; the bird hunter watches the dog; the nonhunter does not watch. When the deer hunter sits down he sits where he can see ahead, and with his back to something. The duck hunter sits where he can see overhead, and behind something. The nonhunter sits where he is comfortable." Beyond animal alertness, Shepard adds, "the hunt brings into play intense emotions and a sense of the mysteries of our existence, a cathartic and mediating transformation." Just so: In wild nature, I find spiritual solace and a cathartic reaffirmation of cosmic sanity that I find nowhere in the made world. To the contrary precisely: Nature is the only antidote to civilization.

And where shall our spirits dwell when wild nature dies?

7

Heartsblood:
Getting Personal

> One cannot help but be in awe when contemplating the mysteries of eternity, of life, of the marvelous structure of reality.
>
> —Albert Einstein

YOU DON'T HAVE TO BE AN EINSTEIN to find mystery, awe, and reassuring reality in nature. Nor, for that matter, do you have to be a hunter.

Hunting is not for everyone. Nor is it available to everyone today. And neither, for that matter, is there any sin in enjoying ethical hunting without a spiritual commitment. As *Bugle* editor Dan Crockett acknowledges: "Choosing to hunt raises no one to higher ground. It merely opens a pathway into a different land." How far that pathway is followed is a private matter.

Yet for those who do choose to walk this ancient trail, hunting is the surest and most natural way into the sacred primal grove of reality, being the same well-worn way we took to becoming human.

Until recently I never thought of myself as a "spiritual person." Certainly not in any ecclesiastic or vernacular sense of the term. In fact, like TV personality Bill Moyers, I'm firmly of a mind that "a lot of religion gives God a bad name." But recently, here in the midst of my middle years, a profound spirituality has bitten deep into my being. To put a name to it, borrowing from two of my betters, I've become an *"Earthiest"* (Edward Abbey), in that "I stand *for* what I stand *on*" (Wendell Berry).

An Earthiest—an original Abbeyism, I believe—is nothing more or less than a pragmatic neo-animist. And animism, I propose, is the ultimate spiritual reality. (What was it Stevie Wonder sang? Something like: "When you believe in things that you don't understand, that's superstition.") While nature—and therefore animism—is palpable, logical, and utterly understandable on any number of levels, "God," by definition, is incomprehensible.

If you can ignore his flamboyant career, his outrageous and often inappropriate outspokenness, and listen to the best of what he has to say, it's hard not to like, or at least grudgingly to respect, Minnesota Governor Jesse Ventura. No matter how devastating his message may be politically, Jesse tries to tell the truth. In this respect, Jesse Ventura is the ultimate contrapolitico in office today—and thus a de facto hero.

For an example specific to our present discussion, in a recent *Playboy* interview, Jesse courageously (or, if you prefer, recklessly) commented that "organized religion is a sham and a crutch for weak-minded people who need strength in numbers."

Physicist Paul Davies restates the issue less spitefully when he says: "To invoke God as a blanket explanation of the unexplained, is to make God the friend of ignorance. If God is to be found, it must surely be through what we discover about the world, not what we fail to discover." In other words— as noted before—organized religion can be a failure of the imagination.

And herein lies the primary fault of all messianic religions: their *absolute self-certainty.* "If a man will begin with certainties," cautions Francis Bacon, "he shall end in doubts; but if he will be content to begin with doubts, he shall end in certainties."

"In our other-worldliness," says C. H. D. Clarke, bringing us back on point, "we have lost the feeling of man's oneness with the Earth, which modern faiths do not deny, but which early chthonic [animistic/Earth-centered/human-humbling] faiths saw most clearly." Indeed. "In indigenous cultures around the world," writes psychotherapist Ralph Metzner, "the natural is regarded as the realm of spirit and the sacred; the natural *is* the spiritual. From this follows an attitude of respect, a desire to maintain a balanced relationship, and an instinctive understanding of the need for considering future generations and the future health of the ecosystem—in short, sustainability. Recognizing and respecting worldviews and spiritual practices different from our own is perhaps the best antidote to the West's fixation on the life-destroying dissociation between spirit and nature."

Commenting somewhat more poetically on this same topic in his weirdly wonderful book *Bone Games,* Rob Schultheis notes that when we lost the hunter/gatherer lifestyle, we lost the deepest spirituality we've ever known: "Something in us died: mojo, obeah, mana, Buddhahood, audacious rapture . . . dead. Dead and buried in an unmarked grave somewhere back there. Our ancestors knew more than we do."

Indeed, according to Paul Shepard, our "primitive" ancestors not only knew more than we do, they were more fully human.

―――――

ACCORDING TO ETHNOGRAPHIC RESEARCH, animism has always been the universal cosmology of unadulterated hunting/gathering peoples worldwide. Likewise, neo-animism—Earthiesm, if you will—plays actively in the cosmologies of all nature hunters whether they acknowledge it or not. As Aldo Leopold points out in his lyrical essay "Goose Music": "Hunting is not merely an acquired taste: the instinct that finds delight in the sight and pursuit of game is bred into the very fiber of the race. . . . The love of hunting is almost a psychological characteristic. A man may not care for golf and still be human. But the man who does not like to see, hunt, photograph, or otherwise outwit birds or animals is hardly normal. He is supercivilized, and I for one do not know how to deal with him."

Unlike the disciples of some sky-bound beliefs, nature hunting and its attendant Earthiesm has never elevated me to the hypnotic heights of "holy rolling" or babbling in tongues, night-tripping with space aliens, or seeing the face of Jesus materialize in a frying tortilla. Yet the intensely emotional spirituality of hunting as personal participation in wild nature—culminating necessarily, at least now and again, in the blood-sealed sacrament of killing and consuming corporeal wildness—often moves me to tears.

And this admission should lead thoughtful iconoclasts to ask: How can you claim to love the same lovely creatures you work so passionately, even joyfully, to kill?

The architect Siegfried Giedion, in a lecture at Harvard, tackled this touchy topic head-on when he asked rhetorically how it was possible "that primeval man both killed and venerated the animal? We have to forget our present attitudes toward the sacred. With primitive men the sacred had a two-fold meaning. It included both the holy and the profane [secular]. Animals were simultaneously objects of adoration, life-giving food, and hunted quarry. This two-fold significance of the animal as object of worship and source of nourishment is an outcome of a mentality which did not confine the sacred to the hereafter. For them the sacred and profane were inseparable."

And for nature hunters, they remain so today: sacred and secular, love and longing, life and death; joyously inseparable.

From my intimate perspective, it is those who stand on the outside, unwilling or unable to comprehend this sacred duality, who lack in spirituality—not those who are consumed by it.

Does the wolf not love the caribou? And does she not undertake her daily hunts with joy? And further, does the caribou, in its deepest phylogenetic heart and soul, not love the wolf? "It is the wolf that keeps the caribou strong," advises a traditional saying, via the predatory mechanisms of selective killing. Which is not to say that predators consciously select marginal or "supernumerary" members of a population for culling. Rather, they catch and kill the prey that's easiest to catch and kill.

It works for the wolf in the short run. It works for the caribou in the long. As Robinson Jeffers observed:

> What but the wolf's tooth whittled so fine
> The fleet limbs of the antelope.

Indeed, without the selective genetic education of ongoing predation, wolf would not be wolf, caribou and antelope would be neither caribou nor antelope, and humans would still be apes. Without predation, nothing would even be.

———

HUMANKIND STARTED DOWN the long winding path to sapience some six million years ago. Across all that gaping void of time, we were gatherers and, ever more, hunters; predatory omnivores; bipedal bears. By comparison, we've been farmers and herders for only the past ten thousand years—less than one percent of our species' tenure by even the most modest of informed estimates.

As both genetics and biology testify, ten thousand years is by no means long enough for a species-specific DNA pattern (genome)—which, according to Shepard's informed speculation, changes about one percent per hundred thousand years—to have even seriously begun adapting to match the radically altered social and physical environment we've wrought for ourselves in that same brief interval.

Were our ancient, instinctive needs for predatory omnivory (and the diet, lifeway, and cosmology it implies) not so deeply etched in our beings—were it merely "something we once did" along the road to becoming human rather than what we *are*—hunting would have long ago been wholly abandoned and forgotten. But such, clearly, is not the case. As Shepard points out: "In defiance of mass culture, tribalism constantly resurfaces." And this in spite of centuries of agricultural civilizations and their various messianic religions' genocidal oppression of animistic tribalism via a concerted effort to bring about what novelist Daniel Quinn (*Ishmael*) calls the "Great Forgetting." Humanity's essential animistic tendencies cannot be genetically obliterated. Yet, in and by urban civilizations, they are sublimated. Thus is the willful destruction of wild nature—which animism would never allow—culturally codified and morally sanctioned in America and throughout the "civilized" world. Moreover, wholesale environmental destruction is the meat of runaway global economics.

Meanwhile, those few among us who would fight to protect wildness or, more daring yet, remain active players *in* wildness—those who clearly have *not* forgotten—are defiled by industrial culture as "tree-huggers, elitists, and troublemakers" in the first instance, "anachronisms, barbarians, and heretics" in the latter.

For my part, so be it. "To embrace the mass religions or ideologies of the present," advises Wyoming meat hunter and poet C. L. Rawlins, "we must first deny what we know in our very bones: how the world works." And how the world works is through an endless sacred cycle of digestion. All things born must die and, one way or another, be consumed. To be or not to be is *not* the question. The question is *when*.

The reality that all flesh feeds on fellow flesh is ineluctable, logically undeniable, even for the strictest of vegans—those whom hunting ethicist Ted Kerasote exposes as hypocritical "fossil-fuel vegetarians."

Merely by purchasing vegetables grown on farms that have replaced former wildlife habitat—veggies that have been chemically fertilized, plowed planted picked packed transported long distances at great expense of fossil fuels and other environmental poisons, then sold in markets that displace yet more wildlife habitat—even the kindest-hearted vegans, exactly like the rest of us, must accept personal responsibility for the deaths of multitudes of living things, large and small: rodents and ground-nesting birds and even deer fawns—the iconic Bambi—cut to bloody shreds by the diesel-belching machinery of agricultural farming . . . while elk and deer and myriad smaller creatures, millions of wild lives annually, are smeared grotesquely across our habitat-devouring railroad tracks and runways and highways by vegetable-transport trains and planes and semitrailer trucks and, sometimes even, vegan-driven SUVs.

And of course we must consider the vegetables themselves—a wholly different tribe from our own, granted, but animate beings no less: lives that die so that we can live. Thus the life/death dynamic persists even in veganism, and just as directly, if somewhat less forthrightly.

To trot out one of Paul Shepard's more colorfully cranky aphorisms: "The human digestive system and physiology cannot be fooled by squeezing a diet from a moral. We are omnivores: our intestines and teeth attest to this fact. . . . Vegetarianism, like creationism, simply reinvents human biology to suit an ideology. There is no phylogenetic felicity in it."

Phylogeny, of course, is the evolutionary history of a species, compressed into a common DNA wiring diagram. Veganism is felicitous to the phylogeny of *no* omnivorous species, jutting like a bent spoke from the great grinding wheel of biological life.

In contrast, nothing could be more in tune with nature, thus more moral, than to follow our omnivorous instincts, needs, and "God-given" talents as hunters, openly and gratefully acknowledging the deaths that go to nourish our lives. In attempting and accomplishing such humility, I propose, a far higher percentage of hunters succeed than do vegans.

A bumper sticker piously proclaims: *If it has a face, I won't eat it.* Yet when we look close enough, *everything* has a face. As traditional Eskimo wisdom professes, "all food consists of souls."

Nor is it merely a question of diet. As an anonymous friend and teacher—let's call her B.D.—countrywoman, vegetarian, philosopher, says of her own northern California rural environs:

> The interplay of life and death is everywhere here: in a post-season fly caught and eaten in a spider web above my desk; in the deer bones, freshly gnawed in the canyon across the stream; in the oak leaves, fallen and now decaying in a mat behind the house. I have been a vegetarian for more than twenty years, which I once thought exempted me from the violence that accompanies the securing of food. But a few weeks of working in the garden my first summer here . . . did away with that comforting illusion. . . . I soon grew uncomfortable with the notion that even a berry might not have a life. Each death is clearly part of sustaining another life, and, just as clearly, my own survival depends on being part of this chain every day in one way or another. Most of the time, I understand this inescapable reality well enough to justify my own role. But sometimes the darkness at the heart of that logic breaks through and I face what seems an intolerable truth. . . . I will never know enough about the profound complexities of life on Earth to be sure that I perform this act—that I kill—with moral certainty. The conviction of my human inadequacy expands within me. And then, somehow, from somewhere, another emotion sweeps over me, and I am enveloped by a sweet and transforming humility, a feeling so unexpected that the experience can only be called a moment of grace. This feeling, which transcends the hunt and yet is utterly rooted in its essence, brings a sense of resolution to the impossible dilemmas with which I have been wrestling. I finally understand that humility is the key. Only through humility can the soul make peace with the terrible necessity of survival.

Viewed in this insightful light, we're led to ask: Who is the more admirable? Is it the spiritual hunter who feels a deep and sincere (neo-animistic) sense of gratitude for the lives she takes? (And what can we call this other than "humility"?) Or is it the man who wears his veganism as a badge of aloofness from his own human nature?

Vegetarianism, perhaps even to the ascetic extreme of veganism, is a valid moral choice—and certainly a healthy choice when measured against our culture's gluttonous consumption of cruelly produced, environmentally destructive, cholesterol-packed and hormone-laden domestic meat. Yet—and here's my beef—vegetarianism is not the *only* valid choice for spiritual and physical health. For vegans to pretend otherwise, believing themselves to be hovering, angel-like, above and beyond the bloody sea of death-dealing life, is self-delusional hypocrisy. Vegans are merely strict vegetarians: they are not angels. (Hitler, you know, was a gentle vegetarian.)

No body rides for free. No thing gets out alive.

And when you get at it right, it's a perfectly sublime arrangement.

Says poet and roadkill hunter Gary Snyder: "To acknowledge that each of us at the table will eventually be part of the meal is not just being 'realistic.' It is allowing the sacred to enter and accepting the sacramental aspect of our shaky temporary personal being."

"Primitive" peoples knew this. Wild animals know this. And so do nature hunters.

The most adamant of hunter-haters—those whose focus is fixed tight on individual animals and the moment while blissfully disregarding the welfare of entire species—are in some essential place moved, I think, by a fearful denial of mortality, especially their own. The adamant animal rights/vegan/antihunter's emotional microview of life perceives death as not merely unnecessary but criminally unjust. It follows that all who directly cause death—predators two-legged, four-legged, winged, or finned—are a priori evil. Especially we two-leggeds, as we alone "have a choice."

The late Cleveland Amory, founder of The Fund for Animals and still the world's best-known animal rights spokesman, articulated this hyper-humanized view in his candid response, in a 1993 issue of *Sierra* magazine, to a hypothetical query: "You have absolute power—now tell us what you'd do to ensure our planet's survival for the next hundred years." Amory's response: "As a starting point I would throw out any government agency or society with the words 'wildlife,' 'game,' 'conservation,' 'federation,' or 'natural resources' in it. All animals will not only not be shot, they will be protected—not only from people but as much as possible from each other. Prey will be separated from predator, and there will be no overpopulation or starvation because all will be controlled by sterilization or implant." (People as well, I would hope.)

The biological impossibility, not to mention the naked cruelty, of such a vision as Amory's is appalling: Wolves, eagles, lions, fish, birds, dogs, cats, insects—let them all eat tofu! Further, given that prey evolved to elude predator, and predator to pursue prey, if the two were separated as Amory proposes both would shrivel physically and emotionally to the lamentable level of zoo inmates—which, in effect, they would have to become in order to keep them away from one another, so innate and powerful is their mutual attraction.

Such hubristic contempt for the unalterable workings and immutable necessity of basic animal metabolism is terrifying. A world without predation is a world without life. We are alive, the elk and I, precisely because we both must die. Our mortality is the lantern globe that contains, shapes, and illuminates the brief, flickering flames of our lives, rendering us numinous. Eternal life, as per the mill-run messianic vision, would not only be spiritually empty, it would also be aesthetically homely and intellectually unbearable. It's the fragility and moment-to-moment ephemerality of our lives that render them sacred; even as the certainty of our deaths—yours, mine, the elk's—makes us ultimate equals.

One of the most significant, if potentially baffling, scenes in Alaskan anthropologist Richard Nelson's award-winning memoir *The Island Within* comes when a native Koyukon hunter voices the classically animistic conundrum: "Remember, each animal knows way more than you do." In addition to its instincts—that "superhuman" ability to interpret and utilize the finest intricacies of landscape, weather, its fellow creatures, and more—what every animal not only knows but instinctively acknowledges is its place in the great web of life. For millions of years, humans knew this too. A few of us still do.

"We be of one blood, ye and I," Mowgli calls in greeting to his fellow jungle creatures—even as they busy themselves (politely off-camera, of course) killing and devouring one another . . . even as many among them would hungrily kill and devour yummy young Mowgli, given the literary license to do so. With Bambi for dessert.

This is the real world. Nature hunting has helped me not only to accept the biological necessity of life-giving death, but to applaud its practicality and embrace it as sacred.

———

EACH SEPTEMBER, after hunting for as long as a month, finally, at some perfectly unpredictable moment, an animal appears—heart-pounding close, uncharacteristically calm, insouciant. In such magical moments, it's understandable that traditional foragers worldwide believe animals sometimes "give themselves" to hunters who respect them. This concept is expressed beautifully in Native American poet Leslie Marmon Silko's transmigratory "Deer Song":

> I will go with you
> because you love me
> while I die.

In words less lyrical yet powerfully poignant in their animistic insight, José Ortega y Gasset suggests that "the hunt is not something which happens to the animal by chance; rather, in the instinctive depths of his nature [the prey] has already foreseen the hunter."

I do not know. Yet here before me now stands the wapiti. Suddenly all the weeks of effort—all the hours of sleep missed, all the miles hiked and mountains climbed, all the rain and hail and cold endured, all the elements that combine to make a true hunt—merge toward a denouement.

Reenacting the essential drama of human history, my universe shrinks to a single hair on the huge auburn chest. Arm and shoulder muscles flex, bending the bow. When all feels right, fingers relax and arrow leaps away.

The elk, unaware of its lurking predator, reacts as if stung by a wasp, bolting off a few steps—then stops and gazes calmly about, flicking its ears at flies. Does it even know?

I know. And like a man too long underwater, I think my chest must explode with the passion of it all. "Please," I whisper, "die fast." As if granting my plea, the great deer sways, stumbles, and falls. Soon comes the susurrous release of a final breath: breath, *anima,* soul; spirit leaping away from flesh.

After waiting and watching for a while, I ease up close and touch the giant deer with my bow—it does not react. Dropping to my knees, I peer into those dark, inscrutable eyes. And in those mirrored orbs is reflected my own fragility, my own impermanence, my own lurking death. To *not* feel such a unity at such a time as this, one would have to be spiritually numb. Though many do not understand, many hunters among them, this *is* a sacred moment.

"After the killing of an animal," Shepard confides, "there is a stillness, when thoughts of life's brevity and preciousness are present."

"Life," says Ortega, "is a terrible conflict, a grandiose and atrocious confluence. Hunting submerges [us] deliberately in that formidable mystery and therefore contains something of religious rite and emotion in which homage is paid to what is divine, transcendent in the laws of Nature."

Or, as a thoughtful friend once observed: "In the moment of the kill, the hunter stands at the intersection of the most profound of opposites—life and death. He knows not only that those opposites are linked—indeed that one becomes the other—but also that his life depends on being part of the transformation, part of the ultimate, mysterious, ongoing communion of all life. There are other ways to experience life's Oneness, but I wonder if this truth is ever so immediate, so palpable, so full of feeling as in the hunter's act."

DISTRACTING ME suddenly from my moody musings, out in the silent woods and not so far away, a loud shrill bugle sounds—followed by the brittle popping and cracking of heavy hooves on deadfall and the bemused, birdlike coalescence chirps of cow and calf elk.

Life flows on. The cows among that little band are already pregnant, or soon will be. And if the winter is hard, there'll be one less elk mouth competing for scarce winter browse . . .

. . . down there in the valleys, where the highways bristle with tan young cyclists on toys worth several times the annual incomes of the Third World workers who made them, and rumble with graying self-styled "outlaws" on Harleys worth more than my house. Down there in the valleys, where a fresh floodtide of urban refugees washes ashore each summer, seeking shelter from the human storm. And many passionate antihunters among them—bulldozing new roads and building new homes and fencing new hobbyhorse pastures; blindly displacing and starving wapiti and other wildlife in a desperate grab for "the good life" . . . even as they destroy the very thing they ostensibly came in search of, even while condemning "my kind" as savage anachronisms.

THE BAND OF ELK moves on, unseen but not unappreciated. When silence returns to the forest, I turn to the bloody task at hand.

Like other natural-born predators, I suffer no sickening sympathy for my prey; no guilt in the killing. As Mary Zeiss Stange has written: "Far from being a mark of moral failure, this absence of guilt feelings suggests a highly developed moral consciousness, in tune with the realities of the life-death-life process of the natural world. The simplistic analogy of hunting to such forms of male aggression as rape and warfare breaks down at precisely this point, where a kinship is perceived between the hunter and the hunted."

No guilt indeed. But my uniquely human *empathy* is gut-churning. Gazing at the gorgeous beast, my eyes cloud with tears, which I accept without shame. And yet at the same time I am positively electrified, buzzing with Ortega's "mystical agitation." And this, too, I accept.

Nor am I alone. Such powerful polarities of emotion are common among authentic hunters, a double-edged metaphor for the contradictions of life itself. Paul Shepard explains: "The successful hunt is a solemn event, and yet it is done in a spirit of joy. It puts modern man for a moment in vital rapport with a universe from which civilization tends to separate him in an illusion of superiority and independence."

After thanking the fallen elk—note, I thank the *animal* for its life—I hone my knife on a piece of stone and begin the gritty work of making meat: unzip heavy hide, open bulging belly, plunge in both arms to the shoulders and struggle by Braille to free a hundred pounds of steaming organs . . . which, exposed like some cosmic crossword puzzle, I ritually inspect and attempt to name, as if performing an inventory of my own inner self. As always, I'm awed by the rock-hard muscularity of the great heavy heart.

Elkheart . . . hart's blood . . . heartsblood . . . warm and wet on trembling hands.

———

BY THE TIME I'M DONE—two hours it takes—darkness is threatening and I barely have strength left to stumble down the mountain and home. Before I go, I divide the quartered carcass between four heavy cotton bags, each bigger than a king-size pillowcase, which I've been lugging in my pack for nearly a month now, hanging them high from sturdy limbs. Thus do I hoard "my" meat from my fellow forest carnivores. They—the bears, coyotes, foxes, eagles, ravens, magpies, and more—will rejoice in the mountain of viscera and marrow-rich bones: the traditional offering.

Tomorrow my hunting buddy Erica Fresquez and I will make two, maybe three, slow trips up and down this mountain, a long and physically brutal day, to backpack out the two hundred pounds of deboned meat and small but artful antlers.

A job of work it is, and I love it all. Like building my own cabin, be it ever so humble; like getting in my winter's wood, all ten cords of it and more; like gathering morel and shaggy mane fungi, berries, nuts, and red clover flowers (for a delicious and restful antioxidant tea). All such ancient, inherently ceremonial chores are *good* work—"karma work" if you will—in that it exercises the spirit as well as the body.

Unlike a city friend's five-year-old son—who recently felt compelled to ask, "Daddy, who killed this chicken we're eating?"—I *know* where the meat on my plate comes from, and at precisely what cost to all concerned. And in perfect parallel, I know as well where the radiant heat that winter-warms my cabin comes from—and at exactly what costs to all: me and my morality; the elk, the tree, the ecology.

Taken with this open-eyed attitude of gratitude, as it is by all authentic hunters, each meal of self-got wild meat, like each hand-split log fed to the woodstove, is at once precious memory, animistic sacrament, and caloric consummation of the great endless round of life-giving death. As the poet Art Goodtimes sums it up:

Spirit leaping, from shape to shape.

———

IN HER INSIGHTFUL INTRODUCTION to the most visually stunning book of 1998, *The Art of Thomas Aquinas Daly,* scholar Cassandra Langer reminds us:

> The hunter's veneration of animals carries over from the Stone Age, through myths, and embodies a deep-rooted symbolism in magic, religion, ritual, prayer and initiation ceremonies. Mortality in this natural setting gives us the full cycle of life, [of] dying, [of] present and past death. . . . Popular myth and reality remind us we have always hunted, and legends of redemption are part of an age-old ritual of creation and destruction. From this point of view, the hunter and the hunted engage in an eternal cycle of energy that feeds the universe. . . . Disguise the matter as we will, there is no question that throughout our history we have identified the spiritual with nature. Rites of initiation teach the lesson of the essential oneness of the individual with the cosmos.

We live in a made world of devastated wildness, monomaniacal materialism, unconscionable consumption, and weirdly *un*natural spiritual paradigms. Amid all this ruin, for those to whom its multimillennial tradition still cries out, the hunt, recalled and pursued with an animistic sense of reverence and humility, remains a meaningful rite of passage from human disunity to natural unity, acknowledging and reconfirming Cassandra Langer's ancient "lesson of the essential oneness of the individual with the cosmos."

And that cosmos, our cosmos, for all palpable, conceivable purposes, is earthbound nature. Indeed, as Abbey points out, this world—of golden sunlight and eager flesh, of life and death and heart-cracking beauty—is the only world we can ever know. And plenty good enough at that, don't you think?

———

OURS IS AN ANCIENT, blood-bonded synergy with wildness, forged across thousands of generations of eating and being eaten by wild animals, thinking and dreaming and praying wild animals, striving through painting, dance, story, drama, song, costume, ritual, and the hunt to *be* wild animals. Thus, in the beginning and for so very long, were our world and worldview zoomorphic.

Then came the Fall, as metaphorically chronicled in Genesis, from relatively paradisaical foraging in a nature-tended Garden of Eden—down, down—into the bottomless pit of sedentary, slave-making agriculture and increasingly concreted civilization. And with that Fall, more than ninety-nine percent of human history, experience, and wisdom were discarded, dogmatically disdained—if not yet wholly lost.

We cannot escape the warning of e. e. cummings: "A world of made is not a world of born." All cultures are made. And ours is made to mindlessly worship technological efficiency: fast, easy, and certain. To transport this urban paradigm into what should be the challenging, meditative, and magically uncertain adventure of the hunt is to trivialize one of life's most profoundly beautiful and—for those blessed and burdened with true hunters' hearts—spiritually nourishing endeavors.

This is modern dominionistic/sport hunting's central problem: a collective failure of the spirit, precipitating the lockstep erosion of internal ethics and external respect that hunting increasingly suffers today.

Our prelapsarian ancestors all were hunters. And the gentle, infinitely sustainable lifeway these "savage" forebears enjoyed for thousands of millennia was informed by the only natural religion earth has ever known—spontaneous, universal, continually adapting and reinventing itself in tribal foraging cultures everywhere. As defined by Richard Nelson, the animism of tribal hunters embraces and unites all of nature as "spiritual, conscious, and subject to rules of respectful behavior."

Recall, if you will, the unimpeachable wisdom of Dersu Uzala. In modern dominionistic/sport hunting, as in modern civilization, this ancient and honorable heritage is not only lost, it's openly mocked. For hunting and civilization to survive—for hunting and civilization to *deserve* to survive—this must change.

If all those posing so poorly as hunters today could only be awakened to the transcendent rewards of *true* hunting—the trip rather than the destination; participation rather than domination—all of hunting's problems would simply disappear, since no hunter would want to take shortcuts. At the same time, all but the most implacable critics of hunting would be disarmed.

And more broadly: If all those posing so poorly as *humans* today could only be awakened to the joyful spiritual rewards of simple, humble, animistic living—what a wonderful world this could be!

As expressed in haiku by that infamous Buddhist hobo, Layman P'ang:

> How wondrous, how mysterious!
> I carry fuel.
> I draw water.

All I've said here—all I've tried to say about the spirituality and thus the poetry of true hunting—is voiced better, far more beautifully and succinctly, by mountain sachem Art Goodtimes in his vibrant ode to love and wildness, "Skinning the Elk."

> "There's a whole lot of life in these animals."
> George nods, almost like a prayer,
> as I hold the hoofed leg steady for the knife,
> mist rising from the gutted belly,
> skin still warm.
>
> Tempered steel peels back
> thick hide,
> fur,
> the dark meat of the interior.
>
> Secret organs slide, steaming,
> into full moonlight
> on the bed of Greenbank's battered pickup.
>
> I can't stop peering
> into the glazed crystal of those antlered eyes,
> two perfect rivets
> welded to the girder of that skeletal moment
> when the bullet's magic
> cut life short.

After the carcass is dressed
and hung from the branches of a cottonwood tree,
I go inside and try to wash my hands—
but the blood won't come off.

There's no mistake.
I am marked for life.
I wear the elk's tattoo
as its meat becomes my meat,
and its blood stains my blood.

Spirit
leaping from shape
to shape.

In the end, we find sacredness only where we seek it. And only *if* we seek it. Authentic hunters, nature hunters, spiritual hunters, seek and find sacredness in aspen grove and piney wood; in mountain meadow and brushy bottom; in cold clear water and stinking elk wallow; and ultimately—necessarily, naturally—in bloodstained blood.

This is the spirituality, the poetry, of hunting. There is no oxymoron here.

———

Postscript: After reading a condensation of this and the preceding chapter, my dear friend John Nichols—the infamous Taos writer (*Milagro Beanfield War*), meat fisher, and gentleman grouse hunter—reacted with concern, to wit: "Don't go ballistic with spiritual *woo-woo-ju-ju* on us, Dave! Always remember that by and large hunting is just good clean atavistic straightforward simple fun and honest impulse of the species, made all the more palatable in the evening afterwards by a few glasses of straight bourbon, the stars, and the smoke from an autumn fire."

Right you are, Johnny. I agree wholeheartedly. But have you considered that what you've just said—re the atavism, the pleasure resulting from indulging an "honest impulse," the fire, the stars, the quiet post-hunt celebration—is *itself* a testament to the innate spirituality of hunting? Relax, *amigo*. We have nothing to fear but the fear of our own natural selves.

The Unnatural Predator?
Taking Aim at Trophy Hunting

The more the merrier. The bigger the better.

—Traditional

SEVERAL YEARS AGO, I received a thoughtful letter from a stranger. The writer explained that she lived in the Washington, D.C., area, was a non-hunter with antihunting leanings, and had a deep concern for the welfare of wildlife. She had recently read a hunting ethics essay I'd written for *Mother Earth News* (for whom I was then western editor) and was responding with a question—a challenge, really. Her concern was that hunters, due to our "passion for seeking out the biggest and best 'trophy' males, exert an unnatural and negative pressure on the genetic well-being of the deer family." (That family, in North America, comprises white-tailed and mule deer, elk, moose, and caribou.) The writer's challenge to me, politely put, was to refute her "unnatural predator" charge against hunters—if I could.

Back then, I was confident I could. Today I'm not so sure. To begin, we're not talking about so simple a matter as the free choice of hunters to kill what they wish. Fact is, federal and state wildlife management agencies dictate what hunters can and cannot kill, when, how, and how many. Thus we have two tough cuds to chew: First, does hunting in fact threaten the long-term weal of wildlife by focusing too heavily on prime ("trophy") males? And second, should the first assumption prove to be true, who's at fault? Hunters, managers, or both?

―――――

IT IS SAID THAT "Nuthin's easy." This is doubly true in wildlife management. At best, designing and implementing socially and biologically prudent hunting regulations is a difficult and dynamic undertaking in this manic changeling world. Further complicating things, wildlife managers have to play by different ground rules on different grounds (and skies and waters) dictated not only by wildlife demographics and biology but by regional culture and politics as well. No wonder, when it comes to "lasting truths" about the evil or benign nature of hunting on the evolutionary welfare of prey species, there's so little agreement among professionals, practitioners, and their various critics.

Still, it's an important question. To answer conclusively is beyond me— a mere hunter, nature addict, campfire philosopher, and backwoods scribbler. Nevertheless, as by now you've likely noticed, I have opinions on almost everything. Accordingly, I'll trot out a few for our mutual mastication.

―――――

WHEN I FIRST RESPONDED to that letter, more than a decade ago, I argued with all good conviction that overall the "problem" of "unnatural selection" by hunters, trophy or otherwise, is neither much of a problem nor all that unnatural. In defense of this claim, I reasoned as follows:

Humans have been hunters since before we were human. Clearly, humans have played an active and ongoing role in at least the recent evolution of many species, predator and prey alike, and continue to do so. On these grounds, humans clearly qualify as "natural" predators—though our behavior of late, via "progress," has been decidedly unnatural.

Our long history as hunters, combined with the reality that it's human nature to go for the gold in any endeavor, largely explains why people— "nonconsumptive" fans of wildlife as well as hunters and fishers—have a built-in attraction to big, ornate antlers. And, too, antlers are among nature's most elaborate, impressive, and thus collectible works of art: the cervid equivalent to butterfly wings, peacock tails, hummingbird tongues, or the rainbow plumage of male painted buntings—both in artful elegance and biological improbability.

Sadly, a minority of hunters have fallen victim to a wrong-headed quest to get their names and priapac egos into one or another sacrosanct trophy record book as often and as prominently as possible. This perverted stripe of hunting—which I refer to (with double entendre) as "head-hunting" to distinguish it from more benign forms of trophy hunting (more about which soon)—is at the same time an ethical, psychological, and (I've come lately to believe) potentially a biological problem.

Yet (I would have said several years ago) even head-hunting skews the prey's evolution only in rare extremes—as when endangered species (the grizzly in southern Canada and the Lower Forty-Eight, tigers in Siberia, rhinos in Africa) are sought out and killed as trophies.

———

IN SEEKING to define trophy hunting, again we are confounded by the "shaky architecture of nomenclature." Not all hunters who set out to bag a trophy personify the Safari Club International brand of ego intoxication. Some merely see it as a heightened challenge. Others have no interest whatsoever in "making the book," yet desire big lovely antlers to hang in their homes as artful reminders of great good times enjoyed in the great outdoors—in Paul Shepard's terms, "mnemonic totems."

Concerning the latter, Shepard notes that hunters have always venerated the prey by carrying home and displaying the most distinctive and tenacious parts of wild animals, honoring and cherishing these anatomical trophies as the natural treasures they are. As a serious antler aficionado myself, I've been there and, to some extent, still am.

Put broadly: Most hunters dream of bagging big bulls and bucks because mature males—whether or not they are in fact "wiser"—certainly are more aloof and secretive and thus more challenging to locate and outwit. And in many hunted species—cervids and pheasants fly first to mind—males are both more ornate and more vocal than females.

Historically our culture has fostered, codified, even mandated this common hunter's longing for big racks. Most states and provinces, struggling to restore wildlife populations following the devastation of turn-of-the-century market hunting, have for decades restricted the kill predominantly to males.

This quantity-over-quality restoration paradigm turns on the biological fact that since one male can impregnate many females, fewer are needed.

And too, as always, there's the negative influence of outdoor media. For reasons of their own (and we know what those reasons are), advertisers of products and services, as well as their sidekicks among commercial hook-and-bullet publishers, editors, and writers, exaggerate the "more is mandatory and bigger is better" theme in hunting (and fishing), grinding the idea into the susceptible minds of readers who've already been brainwashed in that direction by capitalist culture.

Even so, based on decades of hunting and observation of hunters, I remain convinced that most nimrods, while they may dream and even boast before the fact of tagging a well-hung buck or bull, habitually lower their sights to take the first legal animal that comes along, fearful that it may be the last. Backing this assertion is the statistical fact that the overwhelming majority of hunters go their entire "sporting" careers without ever bagging a "book-quality" animal. Even the most modest buck or a doe in hand, after all, provides better eating—and, for those so inclined, bragging rights—than the most magnificent monster still out there in the bush.

In sum I still believe that, as a percentage of the overall hunting population, devoted and talented headhunters are as a drop in the proverbial bucket. But while I once maintained that this tiny minority, *being* a tiny minority, has no long-term negative effects on any legally hunted species . . . now I'm not so sure.

———

ANOTHER THING I long believed (and guardedly still do) is this: If local circumstances ever combine to defeat the scenario I've just laid out, allowing an unhealthy number of prime breeding males to be removed from a given population of a given species, professional wildlife managers will step in to provide a reliable, albeit retroactive, safety net—usually by severely restricting or outright banning the hunting of that species.

In the broader view, by adjusting hunting regulations and the allocation of licenses to constantly fine-tune such variables as bag limit, ratios of males, females, and age classes that can be legally "harvested" in a given season and area, managers have done an admirable job, overall, of increasing wildlife populations while striving to maintain sociobiological (that is, gender and age class) balance within the wildlife populations under their care. This is especially remarkable when we consider that, following the continental wildlife devastation that culminated during the Depression era, managers often had to rebuild from scant or wholly extirpated populations.

For a few examples of America's miraculous success in recovering decimated wildlife populations, the National Shooting Sports Foundation reports: In 1900 the entire United States had fewer than half a million white-tailed deer; by 1996 the species exceeded eighteen million. . . . By the late 1940s, overshooting and other forces, especially habitat destruction, had lowered the worldwide Canada goose population to some 1.2 million birds; by 1996 their numbers had flown up to 3.76 million. . . . Between 1907 and 1996, Rocky Mountain (also called Yellowstone) elk recovered from a low of forty-one thousand to more than eight hundred thousand. . . . By the early 1900s, wild turkey had been reduced to less than one hundred thousand nationwide; the 1996 population exceeded 4.5 million. . . . During the mid-1900s, pronghorn numbers fell to a low of twelve thousand nationwide; by 1996 they had soared to 1.1 million. . . .

And so on, and on. And all of these recovery miracles were politically motivated and overwhelmingly funded—to the tune of eighty percent of total wildlife expenditures in America, or $3 million *per day*—by hunters.

Yet today, in order to keep things going, wildlife management seems mostly a process of patch-patch-patch. Here in Colorado, for a local example, elk management policies long seemed to confirm the charge by antis that wildlife agencies "farm" game species for hunters. That was back when managers labored to provide the best opportunity for the largest number of hunters to bag a bull elk, with next to no regard for the quality of those "trophies" . . . or, it follows, with next to no regard for herd balance and the long-term genetic health of the species. Humans came first, wapiti second. Or so it seemed.

Under that regime, gradually and predictably, annual posthunt counts conducted by the Colorado Division of Wildlife (CDOW) began revealing an increasingly negative trend in herd dynamics: not enough "mature" bulls (generally and generously defined by CDOW as 2.5 years of age or older) in relation to cows and not enough calves per hundred cows. With most older, larger bulls having been culled by hunters who were, after all, obeying the bull-only law, and with the young-adult "branch-antlered" class (2.5-year-olds) dwindling as well, most males that remained were yearling "spikes" (a term referring to their unbranched, thus spikelike, main antler beams).

In a biosocially healthy cervid population, such physically immature and sexually inexperienced young lads would be denied breeding privileges. But with spikes suddenly in the majority, these adolescents were doing most of the breeding—and doing it poorly. Consequently, birth rates fell and calves were born later in the season, with less time to grow and gain calorie-storing, insulating weight before their first winter, prompting a higher winterkill. Something had to change.

After long and considered study, in 1989 CDOW imposed a minimum legal antler size for bull elk. Under this revised management plan, hunters could take only branch-antlered males—4 × 4 (that is, having four antler tines on each main beam) and larger—an indication that an animal has attained at least sexual (if neither social nor genetic) maturity. At the same time, the number of "antlerless"—cow and calf—permits was increased, taking a little pressure off the bulls while helping to improve the bull/cow ratio.

(We have plenty of elk in Colorado—and too many in winter considering the speed with which their critical habitat is being devoured by development. Therefore, the thinning of females and calves—thus reducing wintergrazing pressure—enhances rather than harms the survivability and fitness of the overall population. Herd thinning also reduces human/elk conflicts such as vehicle collisions and agricultural damage—the latter being paid for out of CDOW operational funds that could otherwise be put toward law enforcement, habitat purchases and improvement, and other more democratic and productive ends.)

These new antler restrictions, fortuitously aided by a long string of benevolent winters, have worked wonders. Colorado today has its all-time biggest elk herds, totaling, conservatively, a quarter-million animals. That's more than any other state or Canadian province and roughly a quarter of the entire North American wapiti population. And through it all, most Colorado herds have enjoyed significantly improved, if never quite optimal, biosocial dynamics.

Locally my own observations suggest that at the best of times under this management plan, even many biologically mature bulls, 5 × 5s and 6 × 6s, had little opportunity to breed, since there were enough true patriarchs around to dominate the rut, hoarding all the ladies for themselves—exactly as nature (that is, natural selection) intends.

But those "best of times," they are a'changing. As a new millennium opens, in areas of the state where the terrain is less convoluted than here in the vertical San Juans, where unwise logging has dangerously reduced wildlife cover, and where logging roads allow easier access for motorized humans, elk herd dynamics again have slipped out of whack.

The good news is that CDOW seems to have recognized the problem early on, and is looking hard at how to restore balance. The agency has set a commendable goal of achieving twenty to thirty males per hundred females, averaged across all the state's elk and deer herds, by the year 2005. While this remains far short of the one-to-one gender blend many biologists consider natural and thus optimal, it's nonetheless admirable.

More encouraging yet, for the first time the Colorado Division of Wildlife seems willing to consider such courageous (some will say "outrageous") steps as limiting the total number of hunters, adjusting (shortening) seasons to reduce human disturbance during the peak of the rut, and precision-allocating licenses (through a lottery system) in order to manage on a more localized scale in line with the recommendations of area wildlife managers. Predictably, some of these measures, if implemented, will prove to be an astringent dose for some hunters and outfitters, many of whom will yowl like stepped-on cats. Consequently, the state is setting itself up for a bureaucratic and public-relations nightmare. And more power to them.

Another alternative—more straightforward, in many areas more effective, and far more eco-logical—would be to close some of the thousands of miles of abandoned logging roads slicing across public lands within Colorado (as throughout the West). But that's the USDA Forest Service's jurisdiction, not CDOW's. And, too, the same wildlife and public-lands "stakeholders"—read: politically significant pressure groups—who want bigger herds, and bigger bulls, want easy access even more.

Here, as elsewhere and all too often, special-interest interference won't allow wildlife and public lands management agencies to do as much good as they otherwise could.

OVERALL—in this ecologically oversimplified world we've inherited and continue to erode—elk and other game species are far better off with hunting than without. Tragically, elk and deer no longer live in a pristine, self-regulating environment. To leave them to their own ends—a no-hunting, hands-off approach—would be to doom them to rapid devolution and destruction via overpopulation, overuse of shrinking habitat, endemic and pandemic disease, and starvation-enforced self-domestication . . . exactly as we're seeing among our own bloated, sickly, overpampered, increasingly pathological, devolving species today.

Returning to our topical theme—humans having negative influences on the long-term genetics of the deer family—nothing makes my skin crawl like a photo of some "nonconsumptive," animal rights hero feeding or petting some poor helpless formerly wild animal that has fallen victim to human "protection." For one easily accessible example of this obscenity, note Fund for Animals national director Heidi Prescott scratching the chin of a scraggly white-tailed doe in a photo captioned "Shoot this?" on page 104 of the November 30, 1998, issue of *Time,* in an article about whether or not children should be encouraged to hunt.

My point here is both biological and spiritual: A thousand petting-zoo deer have less evolutionary/genetic value than one truly wild animal, who—being truly wild—if cornered and forced to the decision would rather kick your face in and run for its life, win or lose, than accept alms from a human hand.

When it comes to keeping deer wild—that is, maintaining the *deerness* in deer—I fear I'm among a minority of hunters (and for that matter, Americans) who would enthusiastically endorse the thoughtful restoration of keystone predators to as many public lands as feasible. Further, I would gladly tithe a portion of my own hunting opportunity and wild meat for the almost unknowable privilege of sharing the woods with wolves and grizzly bears. My payment would be the rare feral joy of hearing wild wolves howl, the inimitable *ambiance* of a lurking grizzly presence, and the knowledge that wildness—that is, natural processes and natural order—is alive and well. But for now, that's a hookah-dream.

The current reality is that human overpopulation, cultural anthropocentrism and personal egocentrism, ubiquitous greed, and our ceaseless transmogrification of wildlife habitat into subdivisions, clear-cuts, golf courses, parking lots—even petting zoos—renders widespread restoration of keystone predator species practically, or at least socially, improbable. Or so it seems—with apologies to my friends, fellow hunters and natural men, Michael Soulé and Dave Foreman, who are heart-deep in attempting to "rewild" North America.

(You don't like the argument that hunting is necessary to control prey populations, to limit disease, ecological damage, and tragic conflicts with humans, and to keep evolution alive in the wild? Then join Soulé, Foreman, and their Wildlands Project in lobbying for the restoration of two primary cervid predators: wolves and grizzly bears. The stated goal of The Wildlands Project is "to protect and restore the ecological integrity and native biodiversity of North America through the establishment of a connected system of conservation reserves." Within these conservation reserves, keystone predators would be "repatriated" and allowed to live and function naturally. For more information, write TWP at 1955 W. Grant Road, Suite 148, Tucson, AZ 85745. Or subscribe to their magazine *Wild Earth:* P.O.B. 455, Richmond, VT 05477.)

AN ABSENCE OF PREDATORS in the wild, human or otherwise, leads to all manner of natural imbalance including—as happened in Yellowstone Park following the 1920s extermination of wolves—an ecological skew of medium-sized carnivores, or mesopredators: especially coyotes, foxes, and raccoons. As Michael Soulé and John Terborgh note in their encyclopedic rewilding anthology, *Continental Conservation:* "The most severe impacts of hyperabundant mesopredators and consumers appear in localities where [keystone] predators are absent and hunting and trapping are prohibited."

Coyotes, for the most striking instance, are marginally equipped predators of deer and other large mammals—literally nipping and nibbling such big prey to death—as opposed to the near-instant kills common among bears, wolves, and cougars. Meanwhile, a superfluity of raccoons, skunks, and such increases predation on rabbits and other small mammals (as well as on the eggs and young of small game and songbirds).

A lot of problems arise from a hyperabundance of "consumers." Soulé and Terborgh go on to note that "in large parts of the East, overabundant white-tailed deer are decimating acorn crops and tree seedlings—thereby altering tree recruitment patterns to an alarming degree. Feral pigs in the South are equally destructive to forests and to the wildflowers that contribute 80 percent of the plant diversity of many temperate forests. Over-browsing by ungulates, native and introduced, is so widespread that wildflowers are disappearing even in some of the most solicitously protected old-growth forests."

So for now at least, for an abundance of clear and good biological, ecological, evolutionary, and even aesthetic reasons, we must have hunting. The challenge, thus, is to make it as good and as natural as it can be.

———

WHILE HUMAN HUNTING is an effective and necessary stopgap wildlife management tool, it's rarely perfect. As Tom Beck (among the world's most impeccably honest and objective hunters and wildlife biologists) warns: "Sources of error can occur in each step of the wildlife management process: sampling herd size and composition, estimating number of hunters and hunter success, weather variables, and more. Using hunting as a herd management tool is like plowing the garden with a Rototiller, as opposed to using a hoe to selectively prune."

Moreover, management of hunting and hunters is too often guided—not by true wisdom embodying biology, ethics, intelligence, and compassion—but by special-interest politics and economic "realities." In other words: Business as usual. Yet despite its flaws and awaiting the advent of Abbey's "good news"—the resurrection of a more natural and "perfect" world—hunting is the most effective (and often the *only* effective) wildlife management tool available. And by and large, it works.

To underscore the case that hunting doesn't harm (and in fact benefits) cervid genetics, one could—like so many of hunting's defenders do—begin by citing superficially favorable statistics. Consider:

- Most big game species enjoy higher populations and wider distribution in America today than ever before in this century, even though their habitat has shrunk horribly.

- Mature males can be found, if nowhere in optimal numbers, yet in reasonably healthy numbers in nearly all North American cervid populations.

- There are no obvious signs of genetic weakening among wild North American cervids . . . and a lot of folks are watching.

Reacting to such boasts, critics can reasonably, if only hypothetically, counter that such observations are ephemeral: the long-range effects of contemporary wildlife management policies lie far beyond our myopic field of vision. And in this they are correct.

But then, in turn, critics of these critics can reasonably counter: Because we can't predict the future with absolute certainty, should we just do nothing now?

━━━

RETURNING TO OUR TOPICAL QUESTION—is the human hunter an unnatural and genetically harmful predator of cervids?—observable evidence by and large adds up to a qualified: "We don't think so in general, but in some isolated cases it's verifiably true."

As early as the mid-1920s, a federal biologist named Aldo Leopold summarized an analysis of deer sex ratios in several states by noting that the percentage of females "in all these deer tallies are much larger than would be expected to occur in undisturbed mammal populations. Their abnormality is doubtless in part due to the selective removal of males through the buck laws." In other words: Even before the oxymoronic Great Depression, when the restoration and conservation of wildlife in America was still in its infancy, the prescient Leopold already had recognized that by forcing hunters to kill bucks, game managers were skewing deer herd dynamics.

Leopold goes on to say that "in hunting antlered big game . . . a deterioration in size and apparently in genetic quality has been observed to follow long periods of trophy hunting." Viewing this section of Leopold's *Game Management* (1933) in its broadest context, I'm convinced that Aldo meant to say, or should have said, "long periods of *excessive* trophy hunting."

What constitutes "excessive"? More here, less there, depending on species and local conditions. But overall, from a genetic prospect, it seems logical to say that the culling of prime wild male ungulates by human predators should not exceed the rate at which the same species would be "trophy-hunted" by nonhuman (aka "natural") predators.

The widespread belief that nonhuman predators serve natural selection by eliminating old, young, sick, and other marginal elements from a prey base is, in general, true. For two decades now—through annual trips to Yellowstone and Glacier National Parks, plus my daily observations here in rural Colorado—I've witnessed this truism in action.

This is not to say that only human hunters ever kill prime male cervids. To a predator's eye, the quality that makes for a "trophy" prey animal has naught to do with gender, age, or health per se, but, rather, with the ease of catching and overpowering a highly alert, mobile, and often dangerous meal. *Opportunism:* that's the name of the predatory game. Most often the old, young, sick, "stupid," and otherwise disadvantaged individuals fill this bill. Yet prime male cervids, due to circumstance, sometimes fall into the "easy" category and are sought out and killed by natural predators as well as human hunters. Here's a trio of examples:

- Researchers in Yellowstone Park and elsewhere report that it's common for grizzly bears to specialize in ambushing mature male wapiti during the fall rut when the bulls are distracted by lust and fatigued by the considerable demands of assembling, protecting, and impregnating a large herd of cows. And what's true now has likely always been so.

- Cougars often kill "trophy" mule deer bucks. Except for the weeks prior to and during the late fall rut, big bucks tend to be loners—and it's eas- ier for a cat to sneak up on one bedded deer than to infiltrate the web- work defenses of a doe/fawn herd. And, too, during summer and early fall, mature muley bucks often hie to the same high, rocky, breezy, bug- free slopes where cougars den, placing themselves directly in harm's way.

- Cougars often kill mature deer and elk of both sexes in winter, as do early-spring grizzlies just out of den. In both cases, the big-footed preda- tors take advantage of deep snow to run down their winter-weakened, sharp-hooved, post-holing prey. Here again, it's the prey's situation—not its sex, age, or health—that makes it vulnerable and thus attractive to predators. A robust, seven-hundred-pound bull elk floundering in deep snow can be easier to catch and overpower than even an infant, elderly, or ill animal on dry ground. Just such a drama played itself out on the nearby Southern Ute Indian reservation one recent southwestern Col- orado winter. In this instance, a cougar that was estimated by the tribal wildlife biologist to have weighed no more than a hundred and fifty pounds (judging from its tracks in the snow) ran down and killed a prime 6 × 6 bull elk that weighed at least four times as much—the lithe cat snowshoeing effortlessly across the drifts as the heavy-bodied wapiti sank with every step and finally collapsed from exhaustion, helpless and doomed.

Similar scenes occur more often than most of us know or suspect. And this is good. Natural predation—catch as catch can—is absolutely necessary to maintain a balanced and intact ecology (literally, "household"). Scientifically managed big game hunting, I propose, while sorely incapable of perfectly emulating the subtle species-honing influences of multispecies nonhuman predation, nonetheless has always been, and continues to be, an important piece in the ecological puzzle. And today, through no effort of its own, it's an increasingly essential piece.

Yet as human population pressures and gluttonous consumption put an ever tighter squeeze on wildlife—primarily by forcing animals into ever smaller and ever less wild islands of habitat—there's not only room but a dire need for management improvement. To claim otherwise is doorknob dumb. And part of that improvement demands that wildlife managers continue to work at enhancing their public relations skills—especially the skill of listening to and cooperating with their critics.

As Aldo Leopold noted a lifetime ago, hunters "have been fighting a rear-guard action for the very existence of sports afield. If we continue to regard the issue as a battle, we shall probably continue our retreat. But if we can see the issue as a mutual problem, confronting not only ourselves but also farmers, landowners, and protectionists, and soluble by their mutual cooperation, then a brighter outcome may be anticipated. This hoped-for mutuality of effort cannot become a reality unless game managers know and understand other viewpoints."

The issue has been recently restated, by Stephen R. Kellert and Carter P. Smith, in *Large Mammals of North America:*

> Wildlife management involving large mammals is typically fraught with scientific uncertainty and competing human preferences and values. This policy-making environment renders it difficult for managers to develop biologically sound and socially acceptable policies. Managers must be prepared to address not only questions of a biological nature, but also human dimensions such as prevailing pub-

lic values, attitudes, and the current sociopolitical climate.
. . . A major challenge of contemporary management is recognizing all wildlife values among all relevant stakeholders,
while not compromising the necessity of basing policies on
the best scientific information.

An efficient government agency, like a balanced individual, is an open-
ended, open-minded, wisdom-seeking entity. Even an old dog, if it stays
awake, occasionally acquires a new trick, a new insight, a scrap of new
knowledge, a cumulative epiphany, as we shall see.

9

(Meta)Physician, Heal Thyself!
Personal Encounters with
Trophy Hunting

> A thing is right when it tends to preserve the integrity, stability, and beauty of a biotic community. It is wrong when it tends otherwise.
>
> —Aldo Leopold

CONSIDERING MY NUMERICAL INSIGNIFICANCE as one lonely hunter, just how much personal responsibility should I feel obliged to shoulder for maintaining "the integrity, stability, and beauty" of the biotic communities in which I live and hunt, especially when so few others—hunters, nonhunters, antihunters, politicians, and wildlife managers—seem to give a damn?

This, I suppose, is the real question being asked, and explored, in the following, at times deeply personal, consideration of trophy hunting and related topics.

———

FOR SEVERAL YEARS, some years ago, I was a homespun yet devoted trophy hunter myself. I became obsessed with taking—no, let's tell it like it is:

killing—a really *big* bull elk, if only once, in my annually foreshortening hunting career. A primary motivation was the challenge. Another, I admit without guilt, was the antlers, which I view as art and icons of wildness in one.

In this latter desire, casting about, I found moral support in Paul Shepard and the Canadian bio-ethicist Valerius Geist. In response to my comment that Boone & Crockett style "book" hunting is both morally and biologically untenable, Geist responded that he himself "began as a trophy hunter, but moved away. Recently, however, it dawned on me that to some, trophy hunting is a holy grail. These individuals are ready to sacrifice for wildlife, they enjoy hunting, and normally return home empty-handed, by choice. I respect that."

Shepard, meanwhile, points out that "the trophy hunter is widely ridiculed for his wall covered with the mounted heads of his quarry. Yet even this is part of a tradition of venerating the animal by special attention to the head—one of the oldest continuing customs in human life, whose antecedents have been carbon-dated at fifty thousand years. Such traditions are, except where corrupted, surely a survival of something much deeper in the human spirit than a souvenir of a vainglorious triumph over a defeated animal."

Sadly, for some, it clearly is "vainglorious triumph." But not for all. There are trophy hunters and there are headhunters. When I was pursuing a trophy set of elk antlers, I flattered myself that I fell among the good, or at least the ego-benign. But perhaps I'm not the best judge of myself. After a dozen years of killing only "meat" elk, I had gradually acquired a need to prove, at least to myself, that I could "do it." In pursuit of that questionable quest, for three years I passed up sure arrow shots at cows, calves, and smaller bulls, knowing that the first two requirements of trophy hunting are patience and selectivity. Aiding this resolve was the fact that I managed to kill a deer each elkless year—and in any case road-killed wild meat was readily abundant, especially in winter, at least back then, when it was illegal to collect it and, consequently, car-killed carrion lay rotting everywhere.

(By way of brief self-defense for breaking the law, I plead that the law was a pragmatic measure rather than moral stricture—an attempt to keep poachers caught with fresh wild meat, out-of-season, from claiming they'd picked it up off the road. It was also a wasteful law, punishing the majority for the sins of a tiny minority, and consequently a law that I and many other rural po' folk happily ignored—and which the state eventually rescinded.)

And then, at last, during the fourth year of trying, my "doing it" finally came to pass.

———

LATE SEPTEMBER. Two weeks into the archery season. After two hours of sitting in evening ambush, a blue-black raven appears above me, croaking and cronking excitedly, its raggedy wings moaning with each slow, deep stroke. Having long since learned what this can mean—at least here in this magical mountain place where I live and hunt; at least during the autumn elk rut—I'm hardly surprised when, moments later, the silent forest hops to vocal life with the spirited chirping and mewing of wapiti. From the limb-snapping sounds of it, a substantial herd is headed my way. So deep into the rut we are, surely there's a mature bull in the batch.

With quiet joy and as much grace under pressure as I can muster, consciously breathing slow and deep (helps to keep a body calm), I mentally prepare for the fateful moment.

The chatty herd is approaching from ten o'clock, but still a hundred yards out and obscured by heavy timber. Before the first of them comes into view, I hear a resonant knock nearby to my right, at two on the imaginary clock face. I shift only my eyes—wait several seconds—and then He is here.

The next minutes play out as if I'd scripted them. The 6 × 6 herd bull, having rushed piggishly ahead of his harem, trots directly to the pool and wades right in for a drink.

Over the years, one thing I've noted consistently about elk herd movement is that a senior cow generally sets the lead, the herd following single-file along a well-established trail, with the herd bull bringing up the rear—except when water is the goal, prompting the big bully to dash ahead of the others to drink his fill before the masses can muddy the pool.

This big fellow, filling that bill exactly, gazes briefly about, including a glance in my direction (I'm hidden just thirteen paces from his garage-door chest), but fails to peg my motionless camouflaged form—then sloshes in for a slurp. When the eight-hundred-pound deer lowers his head to drink, his eyes disappear below a thick-trunked aspen blowdown edging the near side of the pool. His vision in my direction is now so totally obscured that I could stand and dance a (quiet) tango and get away with it. The bull, bless him, has gifted me with a close, relaxed, picture-perfect broadside shot opportunity.

I will go with you . . .

Aware that the rest of the noisy herd will arrive any moment and ruin any chance I have—all those eyes, ears, nostrils; all that churning movement—yet strangely calm, I raise my bow and draw, concentrating on a spot low on the chest, just behind the extended left front leg. When that concentration is perfect, the slender spear, seemingly of its own volition, flashes away and buries itself deep in living flesh.

The fatally arrowed animal exits the pool in a literal explosion of water. Running full-tilt, within seconds he's plumb out of sight, though I can still hear him, crashing around in a tangle of aspen deadfall nearby—or maybe that's the rest of the (still unseen) herd in panicked retreat. Hoping to hold the bull nearby, I grab for the cow call dangling from a lanyard around my neck, squeeze out a couple of quiet, calming bleats . . . and lo and behold, the bull reappears, seventy yards away, and comes walking back, casual as you please.

Frantic to end it, I sneak another arrow from my quiver and snap it in place on the bowstring, in preparation for a second shot, should such an opportunity arise.

But it doesn't. The bull stops, stands frozen for a moment (a statue of himself), jabs once at the sky with ivory antlers, crumples to the ground, and is quiet.

———

AFTER YEARS OF WORKING for just such a prize, the killing has taken but a minute. *Just one minute.* Fumbling to light a filthy cigarette—a self-destructive tradition, you bet, but a tradition all the same—I note that I'm quaking like an aspen leaf in a breeze.

It's six in the evening, with two hours of daylight left, as I approach the inert bull. He is lying amidst a snarl of Gambel's oak, just forty-two paces beyond the spring. For a long time I stand and stare, awestruck. Flooded with emotion I am, but no way remorseful. This great beast enjoyed a long (for bull elk) and gloriously free life, evading disease, accident, natural predators, and me for years; and now he has died—fast, clean, and with dignity—*naturally.*

Searching around, I can find no blood trail, only a big sanguine splash where the bull briefly stood, just before going down. Yet when I open his chest, I see that my broadhead—in effect, a spear-pointed, double-edged razor—has sliced through a heavy front rib and gone on to deflate both lungs, severing the aortic arch en route, coming to rest in an off-side rib. No high-tech bullet, wolf's fang, bear's tooth, or speeding Greyhound bus could have done a faster, more humane job of killing.

By dusk I have the horse-sized carcass eviscerated, excoriated, quartered, and wrapped in cheesecloth (one elk requires four "deer bags"), the quarters stacked neatly on the bull's own clean hide (turned inside up) and covered with a nylon poncho to further protect the precious meat as it cools overnight.

LATER—late the same night—my wife Caroline and I enjoy a ritual, at least semisacramental, dinner of garden salad, brown rice, homemade bread, and a hot-fresh hunk of wapiti tenderloin, lugged lovingly home in hand.

Later yet, my dreams are wild, fierce, and gloriously free: warm phylogenetic flashbacks from the cold old Pleistocene.

AT FIRST LIGHT I'm back on the mountain to bone and bag the night-chilled meat. Praise be to the forest gods, no bear or coyotes found it.

By late morning—after hiking back down, then back up, this time with the help (for once!) of a friend with two mules—I'm homebound with meat and antlers. The latter prove too unwieldy to top-load on a packsaddle, and the skittish mules don't like the big pointy things, having never before seen quite the likes, so I carry them out myself. Together, the antlers and massive head must go half a hundredweight, yet they ride on my shoulders light as angel's wings.

Home again (the hunter), I study and admire "my" antlers. Aside from being huge and lovely, they're unique: heavily pearled, sharply ridged, and mildly palmate where tines join main beams, all of which suggests a post-prime potentate. This bull—I'd guess from antler size and conformation, body size, and tooth wear—was at least eight years old. If so, this almost certainly would have been his final fall of active breeding before being upstaged by younger, more vital males. Thinking and believing this eases my nagging conscience.

———

ACROSS THE NEXT FEW DAYS, as word spreads and friends drop by to see for themselves, I'm asked repeatedly if I intend to enter this bull—the first I've ever killed that would qualify—in the Pope & Young Club's bowhunting record book. "No."

Yet, in a deeply personal sense, these great antlers *are* a trophy—a poignant and lasting tribute to the freedom and dignity embodied both in the beast who grew them and in my own humble self as his tenacious and, I like to think, natural predator.

The elk is dead. Long live the wapiti.

And now what?

The meat has been cut, carefully wrapped, and cached in cold storage. For the antlers, I opt for the savage simplicity of a do-it-yourself "European" mount, exactly as adorned the hidebound abodes of our Paleolithic ancestors: a cleaned and bleached skull, antlers of course attached but lower jaw removed, mounted on a wood plaque and hung from my living room wall. The bull's two amber-swirled "ivories"—in fact, atrophied tusks from a previous evolutionary incarnation—each the size of a thimble, will make elegant earrings for my lovely Caroline. As always, the hide, viscera, and skeleton remain in the forest, stacked like a cairn, marking the spot where their owner breathed his last. The scavengers will be overjoyed, Nothing goes to waste.

———

SOON AFTER THE HARD-WON, long-awaited gift of the bull had arrived, I commented to Tom Beck—nature hunter, campfire philosopher, outspoken hunting and wildlife management reformer, and a senior field biologist for the Colorado Division of Wildlife—to the effect that "now I have no reason ever to kill a cow or young bull again. If I can continue to kill just one nice bull every few years, I'll be happy to eat roadkill in between."

My thinking was: It had required four years of hard hunting to kill this first big bull. Thus, by holding out for him, I had saved three "lesser" elk that I surely could have killed otherwise. Since it would likely take me at least another four years to kill a second bull better than the first—and so on for so long as I might last—I'd be exhibiting commendable self-restraint (all that forgone meat!) and a self-serving sort of altruism (as all altruism in fact is).

Or so I thought. But as so often happens in my relationship with the intractable Beck (who his boss at CDOW refers to affectionately and accurately as "an eloquent redneck" and "an intellectual in overalls"), I was in for an education.

"No," said Tom, never an equivocator, "you've got it backwards. Now that you've killed a trophy bull, you needn't ever kill another one. Six-by-six bulls are in short supply. We need those top-dog breeders to ensure the future of elk and democratic elk hunting. If we don't want to see all elk hunting eventually go to limited-draw entry, or worse, we're going to have to abandon the counterproductive paradigm of 'bigger is better.' I'm not saying you should never kill another big bull. I'm just saying you shouldn't *concentrate* on them. Come full around, and return to taking the first legal animal that happens along. If it's a big bull, you win; just don't hold out for one."

Admirably, almost uniquely, free of rack-addiction, Tom walks his talk and rarely lacks for wild meat. But for me (I counter, trying to convince myself, I suppose, as well as Tom), it's a tad more complex.

As much as I love elk meat, I love elk hunting even more. Doing it Tom's way—killing the first legal animal that happens along—more often than not nets premature "success" and, consequently, disappointingly brief hunting seasons. (Granted, this is not a problem for most elk hunters. But blessed with years of experience, living full-time in prime elk country, and having an intimate knowledge of local elk and their haunts, I'm something of a privileged character.) By holding out for a big one, making it harder on myself, I'm maximizing my hunting time.

Although Tom's view is increasingly popular among elk biologists—as it has been for decades among such luminaries as Aldo Leopold, Valerius Geist, C. H. D. Clarke, and A. B. Bubenik—it lacks majority favor among wildlife managers. Certainly that's the case here in Colorado, where CDOW officials boast that under the current "no spike" scheme, we have a statewide average of twenty to twenty-five bulls per hundred cows at the close of hunting season, with five of those twenty being 4 × 4 or larger.

Locally and thus more precisely, CDOW district wildlife manager Cary Carron came up with a winter 1998/1999 helicopter count of 15.4 bulls per hundred cows, shaking down to 10.5 yearlings (spikes), 3.8 two-year-oldsters (three- and four-pointers), and 1.1 adults (5 × 5 and larger). As I commented to Cary, 1.1 mature breeding bulls per hundred cows is a "dismal" ratio. He agreed. But he also cautioned that the bull count, especially in this mountainous, heavily timbered region, is habitually low and necessarily something of a guess.

"When we're flying those counts," said warden Carron, "we don't go out of our way to search for bachelor groups, which tend to winter in heavy timber. It's hard enough to accurately count cow/calf groups from a moving helicopter, and they tend to stay in the open. Because spikes frequently winter with the cows and calves, their numbers tend to be high. It's a big patch of woods out there, millions of acres, and there are plenty of big boys hiding in it. All we can do is guesstimate their actual numbers."

Nor is this new. This same problem was noticed nearly seventy years ago by Aldo Leopold, who observes in his dry but important *Game Management,* regarding deer, that "the greater ease with which does and yearlings are seen probably . . . distorts the figures."

Yet granting even ballpark accuracy to the statewide count, five "mature" bulls per hundred cows (if you want to consider a 4×4 mature, which I don't) is considered dangerously low by Beck and notable others. But then (I toss back at Tom), the status quo management seems to net good calf production and survival—testified to by the fact, you may recall, that Colorado has the largest elk population (and thus the greatest hunter participation and satisfaction) of any state or province.

Tom grants all of this. And then, like a fist in the face, he hits me with his slam-dunk: "Should we be managing wildlife for hunter satisfaction, or for wildlife?"

———

IN LEAGUE WITH THOMAS D. I. BECK, the wildlife management profession is rich in impressive, irrespressible characters. Another was A. B. "Tony" Bubenik.

Across forty-plus years as a wildlife biologist—in fascist-dominated and later Soviet-controlled Eastern Europe before getting his family and himself to Canada—Dr. Bubenik established himself as a world-class authority on cervid ethology: the scientific study of the social organization and behavior of the various deer species. Shortly before his death, writing in *Bugle,* Bubenik offered "An Immodest Proposal" for enhancing elk management. That proposal—"immodest" perhaps, but important—directly addresses a tenacious question: "Are humans unnatural predators?"

English was Tony's sixth language. Consequently, I had the privilege of Americanizing his immodest proposal for *Bugle* and offer this telling outtake:

Proper herd management cannot be achieved under today's norms of killing either (a) trophy-class bulls or (b) the first animal to happen by. As our most important game management tool, hunting regulations should spell out not just the *number* of animals killed per year in a given area, but the *social classes* of those animals as well. In formulating these regulations, wildlife managers should ask the following questions: Is a given elk population living well, or is it suffering social misery? If the latter, what needs to be done to restore social order? That is, how many and which classes of an unbalanced population are potential troublemakers in a sociobiological sense (supernumeraries)? Once these classes are identified, they should be targeted for culling by hunters. . . . When regulations allow hunters to cull certain classes of a population's members in an unscientific, catch-as-catch-can manner, the herd's social mechanisms can become dangerously stressed. This is exactly the fix we're in today.

Escaping that fix, warns Bubenik, won't be easy: "To achieve such a worthwhile goal will require management practices more scientific, flexible, and progressive than those traditionally and currently in use—practices that may never be accepted and implemented without the encouragement and support of a great many selfless, far-sighted hunters."

While Bubenik and Beck profoundly agree that we need to shift emphasis away from killing prime male breeders, Tom's advice to hunters is to kill the first legal animal they see, while Tony feels this is too "haphazard." Who's right? And why?

My arrow lands somewhere in the middle. If, as Bubenik pleads, wildlife agencies could identify, region-by-region, those local supernumeraries in need of culling—and if hunters could also identify them—then hunters should be assigned to take them out. Trouble is, most state wildlife agencies have neither the personnel qualified to make such fine-line determinations as Bubenik's approach requires (based on the Austrian deer management paradigm) nor the money to pull it off.

Similarly, we can hardly expect the average North American hunter—many of whom, *being* average North Americans, can't tell an elk from a moose—to distinguish (through a foggy scope in low light at two-hundred yards on a misty morning) a supernumerary yearling doe or cow from a sociobiologically essential herd matriarch.

In sum, while Tony Bubenik's call is admirable in its goals and precise in its methods, those methods are based on the tightly controlled European hunting model. That dog won't hunt here in sprawling, democratic North America.

Valerius Geist, also well schooled in the European eugenic tradition of "management with the rifle" (*Hege mit der Büchse*), seconds me in this. He cautions that "it requires an experienced, dedicated staff to cull effectively," even on the typically smallish European hunting preserves. On each of these properties—owned by royalty or the rich, or leased by a tight amalgam of modestly privileged characters banded together in a hunt club—a professional gamekeeper oversees each herd. Consequently, these highly trained shepherds—many in fact are skilled research biologists—know each of their animals as if by name, or in fact by name, and know who needs to go and who should stay to keep the local house in order. Armed with such intimate knowledge and omnipotent control, the gamekeepers carefully choreograph each hunt, strictly limiting the kill.

At the same time, the average European hunter (not being an average European) is far better trained and disciplined than the typical American nimrod. Because membership in European hunt clubs is keenly competitive, members hunt under constant threat of being ousted should they screw up in any significant way. It's a real big stick, and it works.

Here in North America—for all the strengths of our democratic hunting tradition (and equal opportunity to hunt a vast public domain is among its greatest strengths) there is a glaring weakness: with millions of hunters taking the field each fall and a majority of them urban-dwelling outdoor dilettantes . . . well, you've got to keep it simple. Moreover, wildlife law enforcement is ubiquitously inadequate.

And therein lies the practical appeal of Tom Beck's "first come, first served" (on the dinner plate) philosophy. Can't get much simpler than that. And even with its built-in haphazardry, it's a huge improvement over the traditional focus on killing mature males while protecting females and supernumeraries. Of course, for Beck's plan to work requires that state wildlife agencies take the first step by limiting buck/bull permits while liberalizing the availability of either-sex and antlerless tags, thus giving those hunters willing to cooperate the chance and encouragement to do so.

––––––

THROUGHOUT THE ENSUING WINTER (and here, so high in the Rockies, winters are wont to ensue), I thought a lot, perhaps too much, about all of this: about my trophy bull and Beck's and Bubenik's and Geist's challenges—to management, to conscientious hunters, and certainly to me. And amidst all the dross of uncertainty, one point of certainty shone through like a diamond in a mud puddle: the biological and ethical need for hunters, wildlife managers, the outdoor industry, and media to outgrow our collective big-rack addiction.

It won't be easy. Even before their earliest hunting experiences, most North American nimrods are schooled to value big bucks and blockbuster bulls over cows and does. Heavy-antlered males—to restate the standard reasoning, rationalization, and truth—are far more elusive and thus present a greater hunting challenge.

Moreover, the allure of big antlers is both natural and considerable. So deeply is rack addiction embedded in the big-game hunting tradition that most hunters of our time, myself likely included, may never be able to abandon the idea that killing a wise old cervid monarch offers the greatest challenge, and thus satisfaction, in hunting.

And, too, there's the undeniable *aesthetic* allure of antlers. Oregon nature hunter Mitchell Caldwell, in a letter to me, explained it like this:

> I fill my home with antlers—wapiti, blacktail, mule deer. Some come from my off-season adventures in the woods (sheds and winterkills). Others I've found at garage sales. One year I attended the antler sale in Jackson, Wyoming [where sheds collected from the adjoining National Elk Refuge are sold for charity] and fell in love with a gargantuan set of wapiti "horns." I paid $300 for them (yes, a small fortune), and they're worth every penny.
>
> My love of antlers as art is one reason I hunt only mature bull elk, but not the only reason. Big bulls are wise, difficult, and filled with honor. My father has gotten pretty

upset with me over the years because I pass up what he calls "perfectly legal elk"; young bulls, cows, calves. But he's getting used to my philosophy. The antlers of my 1994 bull— the only elk I've killed in fifteen years of hunting—are my most highly valued possession.

I suppose an uncharitable feminist (or a post-Freudian psychoanalyst) might say this big-boy bias, so common among hunters yesterday and today, reflects machismo, sexism, immaturity, and perhaps even male-to-male aggression. And in extreme cases, the critic may be right. But since humans never grow antlers and rarely hunt one another for food or trophies, I don't think the comparison washes across the board. Taking a more charitable view of the same prospect, it strikes me that many male hunters sincerely feel that it's less "gentlemanly," thus less "manly," to kill females or "babies"—a burden not borne by the growing ranks of women hunters, I dare trust.

SUCH SUBINTELLECTUAL RAMBLINGS remain largely moot, however, so long as Colorado (for example) continues to sell unlimited over-the-counter bull-only elk licenses and to permit all manner of patently unsporting hunting technology and behavior—industry-proclaimed hunting "traditions" and "rights" that are at odds with honorable, sustainable, *real* hunting. Why should I pass up a hard-earned thirteen-yard bow shot at a beautiful big bull only to have some road-cruising cretin slam-dunk that same animal two months later with a barrage of scope-sighted rifle fire from hundreds of yards away?

"Because," rejoins Beck, donning now his devil's cloak, "by then the rut will be a done deal for another year and that old bull will have had one more season to pass along his superior genes to the next generation of elk."

Well, then, for evolution's sake should I give up boot-powered September bowhunting in favor of driving ranch roads in November with a 30-06 sticking out the window of my pickup? Would that, in the long view, be more ecologically sound and thus more "moral"?

These days, if you care about any thing, no thing is simple.

A FEW YEARS BACK, I attended a public meeting in which a roomful of local hunters and outfitters vigorously lobbied a fellow named George, then chairman of the Colorado Wildlife Commission. All of them were demanding longer this, easier that, more of something else, all of it boiling down to "More for us, less for the other guy!"

At one point, an outwardly calm but increasingly frustrated Chairman George fired back: "Everyone wants *more* from hunting. Isn't anyone willing to sacrifice, even a little, to give something *back* to the resource?"

A gritty silence followed, which I finally broke by offering that if ever it should come to that, and as much as it would diminish my life, I for one would give up elk hunting entirely for the "good of the resource."

Knowing what I know about wildlife dynamics, I knew it would not likely come to that in what remains of my hunting life. But no one else did. And every man in that room (and all in fact were men) stared at me coldly. No one said a word. Even the unflappable George was visibly flapped. What I had said, albeit overtly hypothetical, was blasphemy. Some in that room, no doubt, even considered it un-American.

To AVOID (or at least minimize) hypocrisy, I must admit that while all of this sounds good on paper and in conversation, when I get out there in the elk woods, other forces come instantly into play. With deer and other game, it's not a problem. But with wapiti, bull hunting defines the experience.

In effect, I'm a victim of my senses. During rut each year, bull elk stink in a way I can only describe as disgustingly intoxicating. (Like many of the most enjoyable things in life, it's an acquired taste.) Their haunting, bugled music is nonpareil in the mammalian world. They decorate the mountains with debarked saplings, hoofed-up and pissed-in wallows. They're a whole lot smarter than your average hunter. The older and bigger they are, the more each of these qualities is magnified. And in my experience, their meat is as tender and tasty as any—certainly superior to the flesh of old elk cows, which live on average twice as long as bulls.

So there's bull elk hunting, and there's "just" elk hunting. In the late-October and November rifle seasons, after the rutting is done and the bulls fall silent, it's no big deal. But during the September bow season, the differ-ence is one of magic and magnitudes.

And so it is, no matter what I say, that come September—with the first whiff of bullish pheromones, the first shrill bugle—all I *want* to do is hunt bull.

Yet I try. And with each passing autumn—having grown a little older (thus, I flatter myself, a little less unwise) and finding it a little harder to crawl around in the mountains for a month—I find it somewhat easier to embrace the Beck rule of trophy hunting: Take it as it comes.

Confessions of a Baby Killer: Exploring the Dark Side of Wildness

> The gore of carnivory, the predatory bite, the lethal stealth
> of the parasite, the decimation of wild "babies" by hyenas
> or skuas, the death throes—all bloodshed and butchering
> seem horrible when crudely anthropomorphized.
>
> —Paul Shepard

IT BEGINS BACK IN 1993, the year before I slew my one big bull elk—
described in the previous chapter—while I was still solidly stuck in my own
personal version of trophy-hunting madness. Back then . . .

———

A COUPLE OF HOURS BEFORE DARK, the clarion silence of the aspen grove is
interrupted by a muffled snap of limb, like an elk padding around in house
slippers. A moment later, a smear of brown glides ghostlike through the
trees, just beyond the little pool of spring water I'm watching from an
impromptu hunting "hide"—a stump to sit on, with a bushy Christmas tree

behind. Listening to the approaching footfalls, my heartbeat doubles. Once again, I fit arrow to string.

When the ghostly noisemaker suddenly steps into the open a few seconds later, the adrenaline rush almost knocks me off my log: No elk, this, but the biggest black bear of the nine I've seen this summer. The bruin is only thirty yards out and inbound fast—almost like he's spotted me and doesn't like what he sees. Or smells. Which is not to say the bear is "charging." Black bears rarely do. All the same, there's a big brawny bruin headed straight my way and suddenly my bow and arrow feel profoundly . . . puny.

Besides, I do not wish—ever, for any reason—to kill a bear. (Why not? Because bears remind me of myself, and of my dogs. Because I don't care to eat bear meat. And because—well, it's a "spiritual thing.")

With Frederick Manfred's novel *Lord Grizzly* and the closing scene of the film rendition of Jim Harrison's *Legends of the Fall* flicking through my head, I think fondly of my skinning knife, buried beyond instant reach somewhere in my pack. Likewise, I crave the canister of hot-pepper "bear spray" I always carry in grizzly country but rarely bother with here in Colorado's banally safe San Juans.

Twenty yards . . . fifteen . . . on comes the bear, with seeming ill intent. Lacking a more clever plan, I simply sit and stare. When the beautiful monster hits ten paces, I'll jump up, wave my arms, and shout *Boo!* (I'm serious.)

At a dozen yards, the bruin stops, literally in its tracks, raises its long brown muzzle to taste the evening breeze, cocks its head—big enough to fill a five-gallon bucket—and peers intently uphill, distracted by something invisible to me. A moment more and the bear goes shambling off toward . . . whatever. In instant-replay retrospect, relaxing, I'm sure the bear was not "coming for me," as it almost seemed for a while there. Even grizzlies rarely attack humans unprovoked, and black bears almost never. More likely, I happened to be sitting—perfectly still, perfectly camouflaged, downwind, virtually invisible—where the bear happened to be headed. Just another of the many delicious trailside attractions attendant to patient hunting.

———

My PULSE has barely returned to idle after the excitement of the bear encounter when a distant bugle animates the silent forest. Across the next half hour, the proclamation is repeated at roughly ten-minute intervals, each time a little closer. As an elk hunter and one who truly loves elk, this is a symphony—or, in the scientific lingo of audio expert Dr. Bernie Krause, a "biophony."

In case you've never heard it, the bull elk's bugled call is (with infrequent exceptions) voiced only during the late-August to mid-October mating season. It's simultaneously an invitation to estrous cows, a challenge to competing bulls, and a musical celebration of life, lust, and liberty. Sounding to my ears more like a wavering trumpet blast or a lonely saxophone wail than a bugle, it's distinctly brassy and exciting in an eerie, ethereal way. Beginning with a low, throaty growl (audible only if you're fairly close), the passionate refrain climbs swiftly two or more octaves, briefly sustains, then comes crashing back to ground, concluding with another low growl and intermittently postscripted by a staccato series of nasal, hiccupping grunts, commonly called a "chuckle."

And this, all of it, is what I'm hearing now, on this perfect autumn evening. But gradually my initial elation turns to despair; a pattern is emerging that isn't in the least reassuring for my carnivorous purposes. From the sounds of it, the bull is contouring across a wooded slope well above me. He's headed, I would guess, for Turkey Spring, one of four small pools strung like pearls along this cloistered aspen bench. Ironically, it was exactly there, Turkey Spring, where I sat through four quiet hours and more just last evening . . . to utterly no avail.

I consider trying to entice the bull my way with a challenging bugle or a flirty cow chirp—then reflect on the number of times calling has backfired with these savvy elk, who get annually reeducated by hunters' mistakes, my own included. Nor can I risk a stalk, what with the forest floor a litter of crackling leaves, the evening breeze squirrelly, and my quarry on the move. Stuck with the status quo, at least for now, I settle down to enjoy the concert. Nor am I complaining.

Following an intermission of several minutes, the bugling resumes with redoubled verve, coming alternately from the immediate vicinity of Turkey Spring, northwest of here, then from a ways below, due west. Inexplicably, the bull seems to be jogging back and forth along the hundred-yard length of the spring seep, bugling first from the top, then the bottom. Odder yet, each time he sings from the lower position, he sounds somehow different—owing, I suppose, to different topographical and floral acoustics.

After several more minutes of this, and only when the bull bugles from both places at once, does it finally occur to me that I am hearing not one, but two bulls. The solo has become a duet.

Now a third, loftier voice counterpoints the bugling. It's a raven, circling low between Turkey Spring and me, glinting black as chipped obsidian under the soft September sun and croaking a raspy chant that sounds to my optimistic ears like *Elk! Elk!* I am not being mystical here; this is what the raven would seem to be "saying," as if warning me to prepare for pending action.

Am I "losing it"? Perhaps not. Alaskan anthropologist Richard Nelson describes just such an animistic human/raven interaction in *The Island Within*. Even today, Nelson reports, the Koyukon people of inland Alaska, still a semisubsistence foraging culture, believe ravens sometimes help deserving hunters by guiding them to game and otherwise turning bad luck to good.

In any event, as a practicing neo-animist I opt to practice on this raven—by taking its advice, ignoring my own, and trying to bugle the nearby buglers in. Even if the two bulls clam up and scram, they won't have seen or scented me. And tomorrow is another day—the final day of the annual Colorado bow season; my last day of elk hunting for another interminable year; and my final chance to gather in the couple-hundred pounds of lean organic meat I've come to depend on both nutritionally and spiritually. Last day, tomorrow, but another day all the same.

My initial beckoning bugle—purposefully timid (so as not to intimidate); imperfect, but credible—is answered by a resounding silence. No matter. Having committed myself, I blow a second wavering scale, somewhat more self-confident, and am frankly surprised when the bigger-sounding of the two bulls not only answers, but ups the bet with a five-note postscript chuckle, like the staccato braying of an excited mule. The thrilling war of words unique to elk (and turkey) hunting has begun.

For the next half hour, we trade insults, that bull and I. The smaller elk, meanwhile, has butted out; gone stone quiet.

But as so often happens, the bugling contest soon degenerates to a standoff; my boisterous opponent is all blow and no show . . . as am I? I'm considering a stalk—the distance between us is only a few hundred yards—when a twig snaps nearby. I crank my eyeballs around that way, and catch a dark form slipping through the trees near where the bear first appeared not so long ago. Has that big brown brute circled around for another go at me?

Uncertain and a little nervous, I sit tight and watch.

Half a minute more and a lean, long-legged bull sporting five antler tines per side emerges and comes tiptoeing (cervid hooves in fact are modified toes) down the hill toward me. The pony-sized creature, freshly coated with dark-chocolate wallow mud, looks black as any moose. On he comes, circling around the spring pool, stopping on the opposite side of the same bushy spruce I'm backed into for cover. From he to me is hardly five yards—and there he makes his stand.

Peeking cautiously through the foliage, I see: long ears erect, rut-reddened eyes roving, black nostrils dilated. Only careful camouflage, a benevolent breeze, and dumb-crazy luck preserve my anonymity.

This is hardly the "trophy of a lifetime" I've been holding out for—how long?—three years now. But the season's running short and so is my resolve. His antlers are lovely, he's carrying a winter's worth of juicy meat on those big bones, and under the circumstances he's "trophy" enough for me.

Meanwhile, over yonder at Turkey Spring, the bullish cursing continues.

From the evidence at hand, it's tempting to deduce that these two fellows have conspired to work together—the bigger (sounding) bull keeping me bugling and distracted while his protégé slips in to reconnoiter the situation. Old-school wildlife biologists detest anthropomorphism—the assigning of humanlike thoughts, emotions, or motives to animals. It bothers me too when folks take personification into the fantastical realm of Bambification . . . but this, based on palpable sensory evidence, is different: An elk "conspiracy" clearly seems the case just now.

In any event, the 5 × 5 has stopped in a spot, breath-holding close as he is, that won't allow for a shot unless I rise from my knees and step boldly out from behind the bushy little tree that separates us: impossible. So here I sit—all revved-up and nowhere to go.

For long, electrified minutes—*minutes*—the bull just stands there, mute and statuesque, as I kneel, barely breathing, struggling to ignore the flies buzzing my face, trembling, knees aching, peering squint-eyed through the evergreen boughs, hoping.

Unwelcome relief comes with the next bugle from Turkey Spring. Apparently hearing some subtle message that's wasted on me, my bull grunts—not an alarm bark; more like *"Huh!"*—then turns and strides away, due north, on a bearing that keeps the little spruce directly between us; Murphy's loathsome law.

With nothing to lose, I determine to give it a go anyhow. As the bull walks away, I knee-step a foot to my left, lean out around the tree, cant my bow to two o'clock (so the lower tip will clear the ground), and come to full cock. The bull has stopped at twenty yards—my maximum range—half hidden behind, half protruding from, a leafy scrub of Gambel's oak. Trouble is, it's his wrong half sticking out. I relax my bow arm. A few seconds more and my trophy disappears into a shimmering wall of brush, trees, and blowdown, headed for Turkey Spring (I imagine) to report back to his comrade.

AEONS PASS in somber silence. No more bugles are forthcoming.

So late in the season, this is a substantial disappointment. Yet to have a grizzly-big bear wander close enough to kick my ticker into near-fibrillation, and a mature bull elk at five yards for five minutes and more, and to have been treated to a prolonged bugling concert, all on the same idyllic September evening when (based on recent experience) I had expected no action here at all—I am elated and grateful.

MORNING. The ultimate day of another year's deer/elk archery season . . . in more ways than one? It's my last chance this year, short of foraging roadkill, to collect wild meat. Even so, I've not given up on bagging some big antlers as well. Consequently, I'm back on the job just before sunrise, squatting on a timbered ledge with the black-dark aspen bench below.

Twenty minutes later, with the first eerie glow of dawn, I ease down the slope to a wide spot on a game trail within bow range of Turkey Spring. This is one of my favorite setups. It's similar to hunting from a tree stand in that it offers excellent visibility, favorable wind currents, natural camouflage (sitting in the dirt with a brushy hill just behind me), and a close shot to the spring—yet, my feet (and butt) remain solidly and safely on the ground, where we terrestrial apes are meant to be. Settling in, thrilled to be back, I prepare for another exciting wait.

Across the next couple of hours birds—yellow and lime-green evening grosbeaks, gray and Steller's jays, generic juncos, mountain and pygmy nuthatches, black-capped chickadees, pine siskins, and more—flock to the pool to drink and splash, chittering and chirping gaily. Odd—isn't it?—that birds like to bathe at dawn and dusk, the chilliest times of day. They must know something we do not.

As the sun gradually warms the scene, squirrels emerge to gambol and chatter. Once a pine marten the size of a house cat—the pine squirrel's primary mammalian predator—weasels past in bemused silence, bringing an abrupt end to the squirrels' careless merriment. The air continues to warm, flies buzz, and my eyelids sting and droop. At midmorning, one distant bugle lazes down from far up the mountain, a somnolent pronouncement that seems to say . . . *Yaaawn.*

I'm about to succumb to a nap myself when I'm startled to alertness by the nearby croaking—like a goose with laryngitis, or an immature sandhill crane—of a solo raven. Judging by its raspier voice, this is not the same bird as last evening. And rather than *Elk! Elk!,* this overgrown crow seems to be shouting *Look! Look!*

Obediently (nothing better to do, and nothing to lose), I turn and peer in the direction of the alarm—and by golly! back in the shadowy tangle of spruce and fir directly below the circling bird, I spy the graceful pistoning of four long, slender legs. Big coyote . . . no, a mule deer fawn. Big for its tender age (five months or so), its spots all gone, sleek and chubby. Must have been an early birth; a couple of weeks can make all the difference. And no doe nowhere.

Without hesitation or sound, the little deer glides to the spring, lowers its head, and drinks. Without hesitation or sound—thinking only of meat; really not thinking at all—I draw my bow, aim instinctively, and release.

The turkey-feathered shaft hisses down the slope and strikes its unsuspecting victim with the sound of one hand clapping. As the slender wood missile flicks cleanly through both lungs—a few inches high on the small target, clipping the underside of the spine—the deer reacts by falling like a stone and uttering a vocalization so unexpected and passion-filled that it knocks the wind right out of me, like a boot to the groin. Not a familiar bleat, grunt, snort, or blow, but an almost childlike bawl. Next comes a brittle snap as the protruding arrow breaks under the falling weight.

Silent now, the deer kicks feebly—once, twice—its hooves churning the air, searching blindly for solid ground and the deliverance Mother Earth has offered in the past. A second more and the flailing ceases.

All is deathly quiet.

I am instantly and utterly unzipped. This is not the way I wanted it. This is not the way it's "supposed" to be. Certainly it was a fast, humane kill, the deer going down instantly and the whole bloody drama playing out in a flash of seconds. Yet I am heartsick, filled with an unaccustomed remorse . . . give me back one minute, please, and I'll give back that life.

Hurrying down to my victim, I trip, stumble, fall, and roll, my vision blurred by tears, my balance confused by panic.

Why do I feel this way? The death bawl—that's part of it; no serious hunter wants to cause suffering, or even distress; even if, as I believe it was just now, it's only the startled *Oh no!* variety, signaling more shock than pain. In fact, we are told that a severed spine precludes any possibility of pain. In any event, it's a damned uncomfortable feeling, or should be, when hunters are forced to see and hear, close-up and passionately personal, the sometimes imperfect results of what we do.

Yet there's more to it even than that: a strange, ineffable, deep-seeded sense of guilt and betrayal that I can't quite pull into focus . . . like a far-off whisper.

But the deed is done. Death is irrevocable.

It's a boy-fawn, I note, and the shallow water of the spring is crimson with his blood, as is the ground all around. Flies magically appear and alight on the wide-open eyes, brown and bright and seeming yet alive. Insulted by what I perceive as the blatant disrespect of these lowliest of scavengers—heaping my own guilt onto them?—I shoo away the buzzing bastards and close the sightless orbs. But deer eyes are imbued with a fierceness of freedom even in death, and the lids float back open, as if the animal were re-animating. I close them again, and again they open wide. Having failed in that small courtesy, I slide one palsied hand down the sleek fur of the young buck's neck, and tell him how sorry I am. And in this I am sincere.

Only after the eyes have begun to glaze, erasing the mirage of lingering life, do I whet my knife on a gritty rock and commence the making of meat.

Given my long-practiced skill at field butchery, and the smallness of the deer, it takes only minutes to eviscerate, excoriate, and enshroud the warm pink form in one of the cheesecloth game bags I carry in my hunting pack. Done, I bind the mummylike bundle with a few turns of parachute cord, heave it to my shoulders, and bear it up the slope, where I hang it in the shade of a ponderosa pine.

Finished, I head for home.

LATER, as I approach the cabin, Otis barks excitedly and Caroline appears in the doorway. Seeing the blood on my clothes and hands, she smiles warmly and offers her own peculiar sort of congratulations: "So, you *finally* got your big bull!"

Taking a deep breath, I shake my head in the negative, as if that gesture could shake off the unwanted truth. "No," I say. "I killed Bambi."

My wife shoots me a questioning look; neither approving nor damning, but clearly perplexed.

I struggle to explain. "It just happened . . . last day of the season and no meat in the freezer and you know how tender and delicious venison veal can be . . ."

"So then . . . what's the matter?"

Forcing myself to make eye contact, I admit that I'm not quite sure. "He wasn't the least bit wary when he came in to drink, and he didn't run off and die politely out of sight like arrow-shot animals usually do. I had to watch, which was hellish. And he cried out—in surprise, I think, or fear; maybe even pain. I don't know. What I do know is: I wish to hell I'd let him be."

"We should always have to witness the suffering and deaths we cause," says my best friend of so many years, taking my cold, bloodstained hands warmly in her own.

I suppose we should. And not only hunters: It would be a character-building exercise and a worthwhile reality-check for every last supermarket and restaurant predator among us to watch, now and again, as the cruelly domesticated animals whose flesh we eat and skins we wear are branded, castrated, ear-tagged, and otherwise harassed and abused, then a few months later packed tight as sardines into truck or train car and hauled bawling and anxious to a feedlot where they stand knee-deep in their own muck while gorging on doctored grain before being herded into the charnel house, laid low by a blow from a hammer-gun and dragged off (not always entirely dead) to the cutting floor, where their blood runs deep as a river.

As Paul Shepard makes clear, the animals that provide our meat and leather today, through the process of domestication, no longer are viewed as sacramental. We think of them as sacrificial. If we think of them at all. Which most of us do not.

BY EARLY AFTERNOON I've showered, forced down a light lunch (not very hungry), changed into fresh camo, loaded my freighter pack with a big muslin game bag and other essentials for the chores ahead, and grunted back up the mountain to where the deer hangs mutely in his pink-stained winding sheet. And as soon as I see him—now "it"—the agitation and regret come gushing in anew.

An hour later, with the deer quartered, boned, bagged, strapped to the pack frame—one load will do it—and rehung from a shaded limb, I prepare for my last-evening elk hunt. As my poetical pal Beck observes: "Bambi eats good, but he don't eat long." Winter up this high—eight thousand feet above the distant Pacific—is a long time going. And cold. More meat is needed.

With only hours left in the season, an empty elk tag, and a near-empty freezer, I've decided that my best bet for success is to end it all at Wapiti Spring, where I saw the bear and was pinned down by that luck-charmed five-point just last night. All the activity of the morning at Turkey Spring, amplified by the odor of blood, would warn away inbound elk and might attract the local bear, with whom I have no quarrel.

Arrived, I make myself passably comfortable on my knotty log seat (in hunting, excess comfort leads to lethargy, inattention, and failure), lean back into the familiar Christmas tree, and do my best to blend, body and spirit, into the woodwork.

Gradually, stasis returns to the evening forest: A restful silence interrupted only by the occasional breezy secret, whispered among the quakies.

TWO HOURS PASS, and I'm brooding again about the fawn, when I hear a close, loud *Crack!* of heavy hoof on deadfall. And another. No house slippers this time.

After several anxious minutes of waiting, the animals finally appear: a roan-dark cow trailed close at heel by a big leggy calf of the year. The cow is clearly nervous, though I can't figure why . . . the evening breeze is mellow and steady, flowing downhill from she to me. And as always, I'm fully camouflaged, including gloves and a face net. In getting here, I didn't walk where the two elk now stand, nor have I so much as farted all evening. Even so, the cow approaches the spring with extraordinary caution, stopping often to peer around, mostly (it seems) in my direction. I respond by squinting my eyes, attempting to avoid any sixth-sense visual connection. Perhaps these are local animals who've visited this place often enough to memorize the vegetation patterns, and sense that I don't belong.

But soon enough, thirst overpowers suspicion. The cow moves to the spring, wades knee-deep in, and lowers her coffin-shaped head to drink. I can hear the sucking sounds as she vacuums in water—horselike, through clenched teeth—and see her Adam's apple running up and down, busy as a well-bucket at an oasis.

Either the cow or the calf would be easy meat, just like the fawn this morning. The calf is old enough to fend fine without its mother—elk weaning is a done deal by mid-August and this is late September. Still, especially after this morning, its presence enhances my reluctance to kill its "Mom." And shooting the calf—even though it weighs two-hundred pounds—is unthinkable. No "women or children" this time around. This time it's big bull or no elk at all. That decided, I relax and just enjoy.

As she drinks, every few seconds the jittery cow snaps her head up—acting more like a pronghorn than a wapiti—and peers around, water cascading from black-haired chin. After this unhurried fashion, several minutes pass while the queenly cervid sates her thirst.

Done at last, the cow shoots me a final farewell glare, then stomps away. Her calf follows without drinking. As their heavy, careless steps fade fast into the forest gloaming, I am left alone again with my nagging troubled thoughts.

———

I DIDN'T KILL JUST NOW, I rationalize, because I'm dog-tired from miles of mountain hiking today, plus the emotional and physical effort of converting a living deer to a bundle of meat, which load I have yet to pack home tonight. Too tired to follow a blood trail, should it come to that, by flashlight. Too tired to "field dress" an elk in the dark—a two-hour job of real work that's promptly and absolutely necessary to assure the meat will cool rapidly and not sour: flashlight clamped in teeth, alert for approaching bears, back aching, knife slipping, slicing a fingertip, mixing my blood with elk blood—then, done with that, wobbling home with the deer on my back, unloading . . . then forcing myself up the mountain a *third* time this long day to bone and bag the elk by semi-Braille.

I know *that* drill all too well already, and need no refresher. Not this night. What's more, with a fall-hungry bear lurking about and two bloody-fresh gut piles in the immediate area, I'd have to build a fire and spend what remained of the night guarding my hard-won meat. And the sky looks like rain.

Of course—admit it, Petersen—had the cow or calf been a big old bull, I'd have killed without hesitation. To heck with fatigue, bears, darkness, rain, and common sense.

Why?

———

IRONICALLY, just as I've given up on another year of trophy hunting and am making ready to leave, a husky-voiced bull sings from somewhere up the mountain, not so far away: taunting me, it seems, with the fact that daylight still has a precious few minutes to run and he *might* be headed this way. I reach for my cow call.

But before I can get it to my lips, I'm distracted by the raspy, intermingled voices of two ravens croaking. From the sounds of them, these are the same two birds I heard separately before—last evening and this morning—heralding the appearance first of elk, then deer. But this time they seem to be scolding: *Out! Out!*

Not questioning the logic of it, I obey the ravens' quoth: gathering scattered gear, I walk off into the evening, up the hill and back to the shrouded fawn corpse, embodiment of shapeless guilt.

———

NOR IS THIS THE END. So tenacious and ineffable is my regret at having killed a fawn—a *fawn* for chrissakes—that across the coming weeks I have trouble chewing and swallowing his tender flesh, sweet and creamy as butter. Somewhat removed from the heart of this personal darkness, Caroline has no such qualms.

Gradually, I begin to comprehend the source of my extreme—some have said "absurd"—distress. In fact, I've known it all along. It arises, I'm sure, from an ancient childhood wound, forgotten but not yet healed and likely unhealable. And in our time and culture, this wound is virtually ubiquitous. It's called the Bambi syndrome, and for half a century it has twisted, tortured, and disfigured America's perceptions of wildlife, predation, and hunting.

Including—if you hunt—even your own.

COME THE NEXT SEPTEMBER, I return to my beloved aspen bench and, as detailed earlier, finally kill my "trophy of a lifetime." Again, circumstances force me to witness the dying. But this time my emotions are familiar and comfortable: a quiet elation moderated by empathy, gratitude, and a somber sadness. But nary a shred of guilt or regret. Killing that big old bull seems the most natural thing in the world.

A few days after the fact, I relate all of this to Milt Beens, proprietor of Wildcat Canyon Archery of Durango, Colorado. A fellow "traditionalist," Milt crafts the artful cedar arrows I shoot and, like me, is an enthusiastic campfire philosopher, Milt chides me for feeling guilty about the fawn but not the bull.

"Killing is killing," Milt advises. "Cute little fawn or big bad bull, it makes not a penny's difference. One life is worth exactly the same as another."

You're absolutely right, Milt—from a moral, that is, uniquely human, point of view. But from the long view of amoral evolution, natural selection, and Shepardian "phylogenetic felicity," you're wrong, dead wrong, old friend . . . which, ironically, is why I was wrong in my guilt for killing a fawn.

In fact, Mother Nature proclaims precisely the opposite, demonstrating that in the realms of species survival and evolution, one life *is* worth more than another, thus dictating that . . . *Bambi must die.*

11

The Bambi Syndrome Dismembered: Why *Bambi* (and Bambi) Must Die

Natural mortality hits the young hardest, and to protect the animal that is most likely to die during the winter, while hunting those that are most likely to live, is a rule that puzzles me.

—C. H. D. Clarke

WHEN I PROCLAIM that "*Bambi* (and Bambi) must die," I'm not merely trying to grab your attention. Rather, I'm talking double entendre. I'm talking Hollywood hokum; I'm talking biological imperative . . . and I'm talking childhood memories.

First: "*Bambi* must die" refers to the world-famous Walt Disney movie fantasy—which planted a seed of ill-founded guilt in my young soul, just as it did, quite likely, in yours. The celluloid *Bambi*, I hope to show, is hardly the darling he's widely considered, but a devious little trickster, hurtfully antinature by nature, and long overdue for cultural execution and burial.

Second: "Bambi must die" in real life as well—out there in the fields and woods. Why? Because Bambi—used here as a metaphor for all young creatures of both genders, wild ungulates in particular and deer specifically—is *born* to die. More precisely, a majority of Bambis, it seems clear, are "produced" as food for predators and "intended" by nature to die young. This is not mere personal opinion. This is rock-hard reality, proved out by the fact that most Bambis *do* die within the first few months of their sweet little lives, for which loss, as we'll see, nature compensates.

And this—the bad as well as the benign—the "tragic" deaths as well as the "joyful" births—we must not only accept but come to comprehend as right and good. Like it or not, this is the way it *is*.

While I don't like it much—all the suffering and death inherent to life— I've learned at least to acknowledge natural reality. And not just reality, but the *evolutionary wisdom* it implies. And that's the imperative I am pricking you with when I proclaim that "Bambi must die."

We'll explore the whys and hows of "God's intent for nature" soon enough. But first, let's take on Hollywood.

FOR YEARS, decades even, I've heard, read, and trotted out the term "Bambi syndrome" as if it were some horrid disease known to all. The term—more insidious than infamous—in addition to denoting emotional animal rights pseudo-philosophies connotes the broadly uninformed (or, more accurately, misinformed) views held by many if not most modern Americans toward wildlife, predation, hunting, and animal "babies" (particularly such cuddly herbivores as doe-eyed deer).

Only after the trauma of killing that fawn did I finally became motivated to re-view the megaclassic 1942 Walt Disney animated fanta-drama that started so much trouble for me and so very many others. The first and only time I'd seen *Bambi* was way back in . . . well, no matter. By the time I got around to watching it again, in the summer of 1999, the video I rented was the "55th Anniversary Limited Edition" (itself somewhat dated, having been released in 1996), complete with a vintage documentary on the making of the original film.

THE TRAGEDY of *Bambi* began in the mid-1880s when a young Hungarian chap, having assumed the name Felix Salten—an ethnically self-cleansed revision of the uncomfortably Jewish Siegmund Salzmann—immigrated to

Vienna. There, as a prolific hacker-out of plays, criticism, and at least one pornographic novel (*The Memoirs of Josephine Mutzenbacher*), as well as a bevy of nominally nature-related children's books, Salten prospered.

And among the pastimes the neo-rich Felix came to enjoy, in company with aristocratic neo-friends, was European-style preserve hunting—which he soon came to view as repugnant: a view he subsequently generalized to all hunting. Written in 1924, Salten's *Bambi: A Forest Life* was intended from its inception to instill antihunting feelings in young children as well as the adults who read to them. (Some critics say, additionally, that Salten's *Bambi* is an antifascist allegory with pacific wild nature representing the victimized masses and storm-booted hunters signifying the well-hated Huns.)

Filled with death and suffering, all caused by an enemy known to the animals only as "Him" (capitalized, as if a biblical reference), *Bambi* hit pay-dirt in a strictly stratified culture where hunting had for centuries been jealously reserved to a detested rich and privileged minority.

Meanwhile, away in America, a parallel plot was developing.

———

BEGINNING IN HIS EARLY CHILDHOOD on a Missouri farm, Walt Disney was a lifelong "animal rightist" and (therefore) hunter-hater. The beginning, legend has it, was when Walt's brother Roy stunned a rabbit with a well-placed shot from a BB gun, then administered the coup de grâce by wringing its neck, all of which little Walt witnessed. Receiving the fresh meat gladly, the pragmatic Mother Disney made rabbit stew for dinner—which Walt refused to eat.

Winning English translation in 1928, Felix Salten's *Bambi* soon found its way to America. Meanwhile, Walt Disney, now a young man, had become a film animation entrepreneur. When Disney read *Bambi,* it was instant love—and, likely, money-lust as well. Script work for *Bambi* the animation film began in 1937. Although Disney's product would be a livelier and more hopeful critter than Salten's shadowy Freudian novel, Disney carefully retained Salten's primary conflict: nature is good, humanity is bad.

Writing years later in *Natural History,* bio-anthropologist Matt Cartmill (author of *A View to a Death in the Morning,* which expresses thoughtful antihunting sentiment) notes that "on September 1, 1939, the day that German tanks struck across the Polish border and plunged Europe into World War II, the film's story editor, Perce Pearce, announced that all predators other than *Homo sapiens* had to be excised from the script. 'There's nobody swooping down and eating someone else and their one common enemy is Man. That's the conflict there—and keep it simple.'"

And so, in *Bambi* the flick, we have a benevolent (if occasionally grumpy) old hooter mentoring such favored owl-morsels as squirrels, rabbits, and quail. Only wild grapes and clover flowers are consumed, and not very many of those. Mostly, Disney's childish creatures just sing, play, nap, and baby-talk.

But misanthropy and antihunting sentiment weren't the only values guiding Disney in the "repositioning" of *Bambi* from book to film. As always with this capable capitalist, profit was a prime mover. In earlier projects, the father of Mickey Mouse had quickly learned that profits could be greatly enhanced by manipulating the emotions of filmgoers. In *Bambi,* the massive manipulation of nature for emotional effect, and thus for profit, is propelled to previously unexplored extremes.

Natural History's Cartmill notes: "Once it became clear that sentimentality and outright anthropomorphism would make money, that's the way the [Disney] films went. . . . They deliberately misinform viewers of basic biological facts." Toward this end, Cartmill reports, "animators were told that they could use whatever human expressions they could impose upon the stiff, elongated face of a deer." Consequently, heads become grossly enlarged and set upon tiny bodies. Muzzles are foreshortened to appear more childlike. Eyelashes are drawn streetwalker-long, and the wee critters' pupils are "permanently dilated . . . like those of a renaissance courtesan on belladonna."

Musical manipulation is also forefront: Underscoring the film's alternately elysian and Dantean moods is a Rousseauesque symphonic fantasia. In all, Disney's cartoon magicians achieved an irresistibly infantile look for Bambi, for his love interest Faline, and for their cartoon cohorts—an impersonation further enhanced by the dubbed-in cutesy voices of "real" children.

Leaving no heartstrings unpulled, the opening scene of *Bambi*—whom all the other baby-talking creatures address reverently as "the young prince of the forest"—is an unabashed rehash of the Christmas morning manger scene. "After the admiring animals finish hailing the birth of the young prince and depart," observes Cartmill, "the camera pulls back to show mother and child nestled in a thorny thicket—a tableau that story editor Pearce referred to as 'that madonna picture'—while a remote, godlike father looks down from a heavenly crag."

After re-viewing *Bambi* from the perspective of a nature-wise adult, I agree wholeheartedly with critic Roger Ebert's description of the film as "a parable of sexism, nihilism, and despair, portraying absentee fathers and passive mothers in a world of death and violence." *Bambi*, Ebert concluded, is unfit fare for young and impressionable minds.

Even for impressionable adult minds, even today, *Bambi* carries what critic Cartmill measures as "the force of a sledgehammer . . . despite its pervasive and repellent cuteness."

———

BAMBI—the classic American children's movie of all time, now more widely available, via video rental, than ever before—is a sickeningly oversweetened, unabashedly misanthropic, emotionally manipulative, false-stereotyping, biologically bastardized, profit-mongering piece of Hollywood poopoo—*a big fat lie.*

So what?

So—if the *real* wild creatures of the forest could speak in their own behalf, I suspect they'd be among the loudest and harshest critics of the Bambi Syndrome, knowing better than we how devastating this Disneyfication has been, and continues to be, to their wildness and weal, creating as it does in impressionable minds of all ages a patently and presciently virtual revision of universal peace and love in nature . . . which simply does not exist. To attempt to "manage" nature after such a kindergarten-cartoonish fashion as *Bambi* portrays and fosters—as members of the animal rights movement have professed they would do—would soon spell ecological catastrophe.

I am no smarmy censor. Yet in defense of truth, reality, and the dignity of natural processes, I repeat: *Bambi*—that monstrously unnatural Hollywood propaganda beast—must die.

And so too, I proclaim uneasily, must a significant percentage of his real-life counterparts.

———

IN PURSUING THE HERETICAL premise that "Bambi must die," let's move along—from my deconstruction of our culture's cutest, most profitable, and ultimately most evil little backwoods bastard-child—toward a constructive criticism of modern wildlife management.

Regarding this goal—and for no better reason than to offer, as Monty

Python would have it, "something completely different" by way of format—here's how I pleaded my case, via e-mail, to one of the world's most respected authorities on wild ungulates, British Columbia's Dr. Valerius Geist:

Dear Val:

Here I am with another pesky question—a chain-linked pair of premises, in fact—which I pray you to confirm or refute. First, it occurs to me that nature/evolution "intends" wild infant ungulates to suffer high mortality, with many or most of those young deaths owing to focused predation during the spring calving/fawning season. In my reading and personal observations, it's apparent that winterkill, starvation, disease, parasites, and predation—essentially all the "natural" killers—"gang up" on cervid young. (Of course, in pristine times, the same was true for humans, with nature exerting her own firm population control and saving us lots of problems. But that's another polemic, best saved for another time and place.) Further, this selective "baby killing" continues among wild ungulates, millennium upon century, without their population base being eroded. Herd composition, meanwhile, would seem to be positively influenced by high infant mortality, in ways you know so well and I won't belabor here. In sum, it seems that an "intentional" excess has evolved in wild ungulate reproduction, cervids in particular, with most of this "planned" surplus of infant flesh going to feed predators and scavengers (the latter primarily via winterkill).

I see this scheme in action when I watch local mule deer does produce twin fawns each June, one of which is predictably bear scat within a month. In addition to bears, we have cougars, coyotes, bobcats, lynx, foxes, and eagles, all of them with young of their own to feed. This leads one (this one) to surmise that evolution has deemed twins as the norm for deer so that one fawn—the stronger, quicker, or merely more fortunate one?—can easily escape while its sacrificial sibling occupies the enemy.

Similarly, witness the incredibly and colorfully concentrated predation—by grizzlies, black bears, cougars, coyotes, and now, especially, wolves—on infant elk babes at Yellowstone National Park during the annual mid-May to mid-June calving season. And yet, Yellowstone wapiti thrive.

Finally, as you well know, numerous scientific studies (Schlegel regarding black bear predation on elk calves in Idaho, McCullough re coyote predation on Tule elk in California, et al.) reinforce my own crude observations regarding a "built-in" excess reproduction among cervids, if not my independent interpretations of a cooperative cause-and-effect relationship between cervid reproduction and selective predation. Regarding those interpretations . . .

First question: Do you agree that cervids (and other wild ungulates) have evolved to produce a superfluity of young in response to and "in service of" their coevolutionary predators?

Second question: Assuming the validity of the above, it seems to me that wildlife managers should redesign hunting seasons to put more emphasis on killing fawns and calves (as well as the females that produce them), and compensate by reducing hunting pressure on mature males. This, according to copious evidence and expert opinion, would help balance herd composition while more closely mimicking natural selection, thus helping to ensure the genetic health and continued positive evolution of prey species over the long haul.

I'm sure you're aware that successful studies along this very "baby-killing" line have been conducted in your own lovely Canada, with elk in the East Kootenay region of British Columbia (Demarchi and Wolterson, 1991) and moose in Ontario (Buss and Truman, 1990).

My interest in exploring the turgid topic of cervid infant mortality in the wild—in addition to a desire to publicly dismember the nefarious Bambi Syndrome—is to do what I can to help destroy the ubiquitous and tenacious resistance among North American hunters to killing "women and children." This, in turn, is by way of attempting to help pave the way for the institution of a more precise and innovative management paradigm.

To be credible in any or all of this, I need to substantiate as root fact that high mortality among wild cervid infants is not only natural but necessary—and therefore should be emulated by wildlife managers and hunters.

What do you say, Val?

Awaiting edification, as always, your eager student.

—Dave

Dear David,

Your arguments warm the cockles of my heart! You are presenting almost exactly my point of view. I have argued that no hunter should be legally allowed to shoot an adult deer until he or she has handed in the jaws of ten fawns! I have also proposed that over-the-counter hunting licenses be restricted to calves and fawns—period. Not even "antlerless." (A limited number of doe/cow bull/buck licenses would be available by lottery.) Neither did I make these arguments in jest. I've "wasted" my own precious hunting tags ever so often on fawns to indicate to my students that I mean what I preach. Now to your questions.

The fact is that fawns and calves suffer a high mortality on the way to growing into competent adults. Generally, only a small fraction will succeed. That fraction, on average, is two out of eight or ten, with the latter figures approximating the total number of fawns/calves born to a female cervid across her reproductive life. In cause-and-effect terms, the high birth rate is both the consequence and ultimate cause of the high mortality. That is, over hundreds of thousands of years, females that produced one calf or two fawns per year (as the case may be) were able to maintain their genes in the population. Those that put out fewer young did not.

You need not invoke some cosmic principle that deer are born to feed myriads of predators—although, factually, that's perfectly correct. *Deer flesh makes predators.* Deer, which tend to twin, have suffered, on average, higher predation historically than have species that bear only one young. If you weigh mother and young at birth, however, you may discover that both the mothers of twins (deer) and singles (elk, bison), in terms of fetal weight relative to the mother's body mass, produce exactly the same birth mass.

For an instance, moose and bison cows weigh virtually the same. Bison normally produce one thirty-five-pound calf, while moose generally make two seventeen-pound twins. In the case of elk and mule deer, females bear about 3.5 grams of neonatal mass per kilocalorie of maternal basic metabolic rate. While elk do it in one package (single calf), mule deer prefer two (twin fawns). One can play with that to show that fetal mass at birth reflects predation pressure and adaptation to that pressure.

What is crucial is that most fawns/calves will die. Only the luckiest and best (fittest) will survive. Therefore, either hunters or Mother Nature can take them. By contrast, healthy adult cervids have low mortality rates from predation and winter. Thus, the logical harvesting strategy is to take fawns/calves during the fall hunting seasons, before winter can waste them, compensating with a lowering of the adult kill quota.

Such a scheme [killing more subadults and fewer adults] generates, on average, an older population of females who, because of their age and acquired experience, make much better mothers, producing larger, healthier, and more fawns, while better protecting them against predation. Additionally, being better acquainted with where to feed, older moms produce richer milk in greater quantities, leading to superior body growth and survival of their fawns.

This higher young/lower adult harvest is the management strategy I favor. It's also the strategy used to manage moose in Sweden, with wonderful results.

In short, a population of deer in which females live well into maturity—say, eight to twelve years—will produce (relatively speaking) far more fawns, so that either more hunters can participate, or more fawn permits can be given out per hunter. And wild predators win as well. Moreover, on average, hunters who take fawns rather than adults are harvesting much better quality table meat, leading to a heightened enthusiasm among their spouses for the hunt.

At the same time, by "keeping our fingers off" the bucks, we might get an adult population with a sex ratio of about eighty bucks per hundred does—that is, a natural sex ratio. Now miracles begin to happen: The rut is advanced and shortened. Fawns are born earlier in the spring and across a shorter time. Ergo: a shorter "spread" of vulnerability to predation and more time to grow large before winter, resulting in fewer fawns lost to carnivores and winterkill.

Another benefit of a high mature-buck ratio is that with more big bucks about, more young bucks drop out of the breeding. These youngsters then can save their precious fat stores for winter. Consequently: better buck survival. Additionally, the young bucks will have fewer, if any, rutting-battle wounds to heal. They can, therefore, stick their caloric reserves into improved body and antler growth the following spring. In this way, when not heavily harvested, bucks, relative to females, not only become more numerous, but larger in body and antler mass as well.

In this scenario—more hunting of calves and fawns, less hunting of mature bulls and bucks—hunters soon will start seeing lots of big bucks. If they wait their turn and are drawn by lottery, they will be able to go out, and likely get, a very nice buck. Meanwhile, they are hunting fawns and bringing home the finest venison.

Additionally, if this management scheme is used for mule deer, invading whitetails have little chance of breeding with estrous mule deer does, as those does will virtually always be defended by a big buck, and white-tailed bucks avoid large muley bucks. Thus, management in favor of producing mature bucks reduces the chances of white-tailed deer gradually "taking over" mule deer populations through hybridization, as they currently are doing in parts of the Canadian and American West.

Focusing the annual cervid kill on fawns and calves is a winner in every respect.

Sincerely, Val Geist

A curious concept, eh? Enhance trophy hunting by *not* trophy hunting!

The same day I "spoke" with Valerius Geist via e-mail, I ran my "Bambi must die" hypothesis by the critical ears of my biologist buddy Beck. Here is Tom's spontaneous response (offered without knowledge of what Geist had to say):

Most of each year's crop of fawns and calves will die before they see a second spring, and the primary killer is winter. Therefore, if you hunt calves and fawns in the fall, you're hardly affecting the infant-mortality dynamics at all. What we're talking about here is compensatory mortality—the fact that most cervid young are going to die within a few months anyhow, making them biological supernumeraries—so we might as well hunt them, then compensate by reducing the harvest of genetically essential adult males. In this way, without reducing herd numbers, we can improve both age and gender ratios, more closely imitating the conditions of natural selection.

And in all age categories, with both deer and elk, we need to be killing more females and fewer males.

Dale McCullough, who's done extensive herd-density research with mule and blacktail deer, presents a clear and convincing argument that we don't want the maximum number of deer on a given range (as we often get with current management paradigms favoring buck-only hunting). Say you have a range that can "support" a thousand deer. With buck-only hunting, you'll eventually wind up with a radically skewed sex ratio, something like 950 does to 50 bucks. What's more, with the range being browsed at maximum carrying capacity, both flora and fauna are stressed.

To escape this trap, you have to reduce the number of does so that the survivors will have better nutrition, leading to more and bigger fawns, along with a higher percentage of male fawns. Although we don't understand the precise dynamics of it, well-nourished does tend to produce more male fawns than females. Additionally, with a healthier range operating at less than full carrying capacity, you get a higher first-winter survival rate for fawns. This is a very compelling argument for hunting females and young instead of mature males—and why I almost never kill a mature buck or bull.

Even should we stop all hunting, as the animal rights folks would like, the ratio of male to female cervids will never return to normal and healthy unless a lot of females are somehow removed from the population. A primary reason for this is that every time you get a hard winter, the males—being nutritionally stressed and maybe wounded from the fall rut—are the first to die.

Therefore, the very concept of "carrying capacity," as interpreted by livestock ranchers and most wildlife managers, is misleading and harmful. You obtain optimum herd balance and health when a range is utilized at *below* its ecological carrying capacity. Rather than *maximum* carrying capacity, we should be managing for *optimal* carrying capacity. And the only way we can achieve that—to reduce a bulging wild ungulate population—is to emphasize the hunting of does and fawns.

The genetic imperative for all life is to reproduce itself. Under relentless predatory pressure, evolution naturally favors the gene strains of those individuals who produce larger numbers of young. Were it not for countless millennia spent living under heavy fawn-predation pressure, deer might never have evolved to the present norm of having twins. But because, genetically, they "know" they'll be losing most of their young, they have "learned" to produce more—an excess large enough to compensate for predation, winterkill, and other baby killers.

For all these reasons and more, in most situations, were it possible to make the distinction, the hunter's primary target—from biological, ecological, and evolutionary points of view and thus from an *ethical* perspective as well—should be female fawns and calves.

Nor is that distinction—fawns and calves versus mature deer and elk—difficult to make, even through a rifle scope at a distance. With just a little added experience, patience, and will, hunters can also learn to distinguish between yearling and adult animals—based on body size and facial configuration, with or without antlers. This is a critical element of basic hunter's woodcraft . . . alas, a largely lost art in these days of virtual living.

———

IN AN UNPUBLISHED MANUSCRIPT titled "Quality Deer and Trophy Management in Europe," Valerius Geist agrees—in general, if not in every particular—with Tom Beck's approach to revamping the American trophy-hunting paradigm. Moreover, he names others who feel likewise. In this academically detailed paper, in addition to demonstrating his encyclopedic knowledge of the centuries-old history of preserve hunting and so-called quality deer management (QDM) in Europe, Geist opens with applause for the precepts of QDM as recently popularized here in North America:

> The central tenets of North American quality deer management, as articulated by Wegner (1993) and Marchinton et al. (1993) [see bibliography for citations], are that current white-tailed deer management should move from a heavy kill of predominantly young bucks, and an inadequate harvest of antlerless deer, to one of sparing young bucks, reducing the number of adult does, lowering total deer density, and improving nutrition. The aim is to produce not merely large adult (quality) bucks, but quality does and fawns, quality habitat, quality hunting experiences, and quality hunters. The call is out to improve self-discipline in hunters, so that they become selective in taking deer—most notably, in forgoing the killing of young bucks. There is also the important admonition that hunters should view deer as part of a greater universe to which we return for the experience of belonging. Deer hunting and management, it is suggested, are in part a matter of reciprocity, a matter of giving to and taking from deer.

The weakness of QDM in Europe—and its potential downfall here in North America is, according to Geist, that it "glorifies trophy bucks. That's why I am not wholly with it. And even though I am in agreement that there needs to be an antlerless cull, I am far more emphatic than QDM advocates in arguing for increased hunting pressure on young of the year."

Indeed, if Responsive Management's survey of Arkansas hunters is representative of hunters across North America—which, frankly, given regional cultural differences, I doubt—then Geist's concerns may be justified: American QDM may take the same wrong turn toward trophy production that most of Europe took so long ago. According to RM, when Arkansas hunters were asked what "Quality Deer Management" meant to them, forty-eight percent "offered meanings related to increased size of deer/bigger bucks." Another six percent equated quality with increased quantity of deer. Only twenty-eight percent of licensed Arkansas hunters correctly correlated QDM with "increased overall health of the deer herd."

Looking hopefully toward the essential goal of overall health for North America's deer herds, Val Geist, in concluding his treatise on European QDM, says "the best way to insure genetic diversity of a closely managed local cervid population is to apply the Norwegian model, in which the annual reduction kill is primarily focused on the young of the year, leaving adults to grow, develop, reproduce, and age. The small adult annual kill should not be selective as to age, size, or sex."

In other words: Children first, then take 'em as they come.

And yet it's that "closely managed" proviso that makes QDM difficult to apply across North America's vast public domain with its millions of hunters.

In sum, scientific consensus is growing in favor of having hunters exact a larger tithe of Bambis and Falines for the sake of cervid posterity. Wolves can do it. Winter can do it. Hunters can do it. Which would you prefer?

———

REINFORCING THE TOUGH-LOVE wildlife management arguments expressed here by Geist, Beck, Clarke, Leopold, Bubenik, Petersen, et al. is a "popular" (as opposed to scientific) article that appeared awhile back in *Sports Afield* (shortly before that venerable hook-and-bullet standard found political correctness and temporary bottom-line salvation by becoming yet another *Outside* clone).

In that piece, Mary Zeiss Stange recalls a meat hunt she and her husband, both then graduate students and poor, undertook the day before Thanksgiving in the "record-breaking cold" winter of 1985, out on the frozen, wind-scoured plains of eastern Montana. Stange's story, "In the Snow Queen's Palace," is the most subtly effective assault on the Bambi Syndrome I've seen.

Early in their hunt, she writes, "after several hundred yards of crunching through ice-encrusted grass, we sprinted along a fenceline toward the creek bed. Skidding down an embankment . . . we were suddenly brought up short by what lay ahead: a little fawn, curled as if asleep, snugly nestled in the snow drifted against a fencepost. The tiny deer was frozen solid. . . . Doug knelt and stroked the fawn affectionately, as one might a cat or dog found napping by the woodstove. . . . A little death like this could not go unmarked, unmourned."

Farther on, the couple encounters a group of deer clustered in a patch of woods, too cold and winter-weak even to run away as the hunters approach. These pitiful creatures were, Stange felt, "trapped by their instinct to survive. In such a situation, shooting was impossible." Exchanging "a quick glance and a wordless nod," she and her husband turned and walked away. Finally, as they themselves are about to freeze and their hunt nears its end, the couple . . .

> spotted some deer at the far end of the field, several hundred yards off. No more than indistinct shapes, they appeared and disappeared, phantoms in the now thinly falling snow. With the wind in our favor, and the snow as much camouflage for us as for them, we proceeded along the fence until we were perhaps two hundred yards away. Steadying my 30-06 on a fencepost, I focused the scope on a doe, another doe, a fawn, another fawn. Teeth chattering and my right hand burning with cold . . . I thought quickly about the fawn we had seen curled peacefully in the snow, about those deer paralyzed for sheer survival in the trees. "I'm taking the one farthest to the left," I whispered to Doug, who was also aiming his rifle. I placed the crosshairs for a heart shot, and fired. His shot came an instant later. We had killed the two fawns.

While the author knows her biology and relays it in what I view as an artfully subtle manner, some *Sports Afield* sports failed to "get it." Predictably, the magazine's letters column soon bristled with outraged missives, accusing Stange not only of being a baby-killer but of glorifying her heartlessness in print. What these critics missed, being ignorant of the implied biological wisdom, is that had the hunters killed the two does, both fawns would surely have died as well, given their inexperience in finding food and shelter and the severity of the winter. Even had the hunters spared both does and fawns and killed nothing, statistical probabilities predict that the youngsters would have perished, as had that poor little Popsickle by the fencepost, which could have been a sibling.

Certainly the hunters could have killed one doe-fawn pair, thus keeping together the other "family unit." But then, the one spared fawn would likely perish in any event, leaving only one doe to reproduce herself come spring. By killing both fawns and sparing both does, the hunters had gotten their meat without harming either the postwinter population or the reproductive potential of the local herd.

Wolves could not have done it better, and likely not as well.

———————

EXTENDING THIS DRAMA back into the mists of antiquity, there can be no logical doubt that hominid predators have been bopping Bambi forever.

This was brought home to me last May while enjoying an evening walk in the woods with Otis the dog. As we slipped quietly along a wooded game trail bisecting a profusely vegetated aspen grove—like walking through a tunnel of shivering green—an elk cow rose from her bed, just twenty yards ahead. We stopped short (from long practice in the presence of large wildlife, Otis knows the drill), and the young cow just stared.

Reckoning—from the time of year, the place, and her oddly courageous behavior—that the cow had a calf stashed nearby, and not wishing to disturb them further, we turned and began to backtrack. Within a few steps, Otis's sharp nose, or maybe his eyes, caught something we'd both missed coming in: a tiny spotted Bambi elk, curled beside a log, a few feet off the trail. Too young to run away and instinctively paralyzed by what biologists call the "hider strategy," the little thing just lay there, moving only its big brown eyes, hoping (I imagined, personifying) for the best. I quietly called Otis back—he only wanted to sniff the furry bundle, but even that was too much—and we made ourselves scarce.

Later I recalled that certain ceremonies once practiced by certain American Indian tribes mandated that the unblemished hides—unpunctured by arrow, spear, atlatl, or knife—of infant cervids or pronghorns be used in the manufacture of certain sacred costumes. Without shooting them, I long wondered, how did the Indians acquire those tiny pelts? Only after practically stepping on that elk calf did it all come clear: If I had wanted that calf, it would have been as simple a matter as whacking it on the head with a club or rock; or, as Indians in some cases did (following religious prescription), strangling it to death.

Likewise, long before prehumans developed the tools and skills of true hunters, they doubtless included the capture and killing of helpless infant animals in their foraging/scavenging strategy—just as surviving foragers still do today. (Just as all natural predators still do, and we supermarket predators as well, by cruel proxy, when we buy veal or lamb.)

By way of hammering yet another nail into woebegone Bambi's casket, researchers Raymond A. Demarchi and Anna J. Wolterson, in a paper titled "Results of Special Calf-Only Hunting Seasons in the East Kootenay Region of British Columbia," observe:

> The strategy of maximizing calf harvest may be difficult for some hunters to accept. However, it is our experience that if [we explain] that calf harvests form the basis of both the cattle ranching industry and natural [wild] ungulate population regulation, most hunters will agree with the concept. Hunters in the East Kootenay have demonstrated a willingness to harvest calves when directed. [More than] nine years of experience . . . have proven that directed calf elk hunting is an effective management regime in regulating elk population structure and numbers while increasing recreational hunting opportunities.

After running this chapter under the alert eyes of friend and adviser Dr. Florence R. Shepard (whom her M.D. daughter dubs an "existential phenomenological mythopoetic reconceptualist") Flo responded: "Surely, you aren't trying to propose a whole new wildlife/hunting management policy . . . or are you? Do you, as a self-professed neo-animist, actively endorse the 'management' of wildlife and wildlands . . . or are you merely trying to show that current policies are ill informed, out-of-synch with contemporary ecological knowledge, and driven by antiquated social 'values' rather than ecological wisdom?"

Well, Flo, I'm not really sure. I certainly don't have all the answers. All we can do is try our best. But knowing what I do (think I) know—and borrowing heavily from those who know infinitely more—at the least, I shall cry no more forever when Bambi bites the dust. . . . I hope.

12

Why I Bowhunt: Confessions of a Pleistocene Throwback

> The contemporary hunting mainstream [is] an arena that
> has become so mechanized that you may as well stay home
> and fiddle with the Internet for all the good it does your
> soul.
>
> —Jim Harrison

I'M ALWAYS SURPRISED when someone cares enough to ask why I choose to
hunt with a bow and arrows—and more, why I go "traditional," a term
embracing simple longbows (a straight piece of wood or a wood-and-fiber-
glass laminate) and recurves (essentially a longbow with its tips bent forward
to provide enhanced spring, or cast), together known as "stickbows"—while
eschewing such high-tech, mechanically assisted missile launchers as crossbows and compounds. (The latter are drawn and held by muscle power, but
they use pulleys, cams, and a sophisticated cable arrangement to lessen the
amount of muscle required.)

The short answer is this: Traditional archery is to hunting as fly casting is to fishing—each is the quietest, most artful, most personal and engaging expression of its genre.

To go farther, dig deeper, is no easy task and risky to boot. Trying to say why you prefer bows over rifles, stickbows over compounds, high-powered rifles over muzzle-loaders, or redheads over blondes (or blonds)—explaining any personal preference—implies a belittling of the options rejected and, by extension, an insult to those who embrace those options. To truly speak your mind—to speak your mind truly—is to offer your neck to the chopping block of rebuttal. Maybe that's why so few people do it.

But me . . . well, I'm a charter member of the First Amendment fan club, with skin as thick as my skull. So here we go.

———

FIRST, THE BIG PICTURE: According to research conducted by Responsive Management, in 1986 only nineteen percent of hunters reported over bowhunting. By 1991 that share had risen to thirty-three percent. RM's most recent survey, conducted in 1996, showed that the percentage of hunters using a bow at least part of the time had more than doubled since 1986, to forty-five percent.

When RM asked why, fifty-eight percent of bowhunters responded "challenge and skill requirements." Detailed responses shake down like this: lengthens overall hunting season, twenty-four percent; less crowded, thirteen percent; to hunt during earlier seasons, eleven percent; change of pace, six percent; peace and quiet, four percent; because it's fun, three percent; and three percent hadn't a clue. While forty-seven percent of male hunters hunted at least part-time with a bow, only twelve percent of women hunters were archers.

———

I LIKE BOWHUNTING because it's harder, more challenging, more instructive and rewarding. Ditto my preference for traditional gear—two sticks and a string—over cams, wheels, pulleys, overdraws, stabilizers, sights, range-finders, trigger-releases, and all the other *stuff* that so clutters and complicates bowhunting today. It's the old military acronym KISS: Keep it simple, stupid. And what was it Ortega y Gasset offered along this line? Something about the dangers of confusing the hunt itself with that which merely has to do with hunting.

This is not to assert (as he moves swiftly if clumsily to beat his critics to the pass) that all archers are more skillful, dedicated, or heartful hunters than all gunners, or that stickbow benders are in all events superior to pulley pullers. Some of the best hunters I've known have never owned a bow. And some of the sloppiest, most unethical, and hurtful hunter behavior I've been cursed to witness was perpetrated by men (always, so far, men) with bows in their hands. There's a bit of yin in every yang.

Moreover, many hunters, myself only rarely included, live on both sides of this coin, toting today a bow, tomorrow a rifle or shotgun. Of such distinctions I am acutely aware.

All I'm saying is that choosing to hunt with a short-range, single-shot, muscle-powered tool such as longbow or recurve—knowing that your chances of bringing home the backstraps (equivalent to the large side of a T-bone) decrease in lockstep with the speed and reach of the projectiles you shoot—such a relaxed, purposefully inefficient approach to "success" suggests an extraordinary passion for the hunt bound up with an admirable deemphasis on the kill.

By "inefficient," I mean only to poke fun at the term "efficient" as commonly applied to hunting tools designed to make killing easier and more certain, with less skill and effort. In this regard archery, especially traditional archery, is gloriously "inefficient." I am *not* saying—nor can you convince me—that the modern bow and broadhead-tipped arrow, in calm, skilled hands, is anything but an efficient and humane hunting tool. Arguably death by broadheaded arrow is among the least painful and thus the most humane ways for an animal—including, should it come to that, a person—to die.

Writing in *Traditional Bowhunter,* veteran bowyer and hunter Dan Quillian points out that "in almost every talk he makes, Wayne Pacelle, formerly of the Fund for Animals, includes the words, 'Bowhunting is the cruelest sport of all.'" Arguing to the contrary, Quillian cites an informal study conducted by a Ph.D. candidate in wildlife management who did a stop-frame examination of "every video he could find that contained pictures of deer being hit by an arrow. . . . In every case, the deer were in motion *before* the arrow hit." While this certainly doesn't rule out pain when the arrow does hit, it suggests that it's the snap of the bowstring, not the pain of impact, that causes bow-shot animals to duck and flee, when they even bother.

Quillian goes on to tell of two longbow hunters who recently reported "shooting completely through the chest cavity of deer and having the arrows hit the ground on the other side. The deer's reactions in both instances were to turn and look at the arrows that had gone through them. . . . Both deer died within a few yards of where they were shot, but they never spooked."

Like myself and many other bowhunters I've spoken with across the years, Quillian has often watched in amazement as bow-shot cervids, after an initial flinch or look around, go back to calmly feeding!

Continuing his informal research, Quillian next "started collecting stories of people who had been hit with broadhead arrows" and survived. He found five.

The first victim fell from a tree and "drove an arrow almost dead center through his thigh until the point pushed out the skin on the other side." He felt no pain.

The second didn't even know he'd been accidentally shot by a hidden hunter until he "unconsciously rubbed his hip and felt something wet and sticky. He looked at his hand and it was covered with blood. When he looked down and back he saw an arrow protruding from his rear end." Which is to say: He felt no pain.

"In the third case," writes Quillian, "the archer was also shot in the hip and his story is almost the same as the one above." He felt no pain.

The fourth victim was likewise accidentally shot, through an ample gut, the arrow "protruding from both sides of his jacket at his belt line. The first thought was that the arrow had gone only through his clothes until he unzipped his jacket and saw that the arrow had completely penetrated his body, extending from side to side. He felt no pain."

Thigh, hip, guts. . . . Would things be different, Quillian wondered, if an arrow hit bone? Not for the fifth victim, who "felt a blow from behind on his shoulder, as if someone had struck him with a fist. A broadhead arrow had buried itself in his scapula. Beyond the initial blow, he described no pain."

Dan Quillian concludes his refutation of Wayne Pacelle's unfounded accusation by citing a medical text's "pain scale." The upper slots were filled by "poisonous snake bite, passing a kidney stone and having a baby. . . . At the lower end of the scale were such things as puncture wounds, cuts, and some cancers."

In my own case, the only personal evidence I can offer (or ever wish to offer) is slicing my finger with a razor and stepping on a shard of broken glass. In neither event was I aware I was cut until I saw the copious blood.

Considering the various grisly and often excruciatingly slow "natural" deaths wild animals commonly suffer—together with a general consensus among researchers that even the most intelligent animals don't experience pain to the degree humans do, along with Quillian's arguments, my own observations, and more—I feel confident in saying that far from being "the cruelest sport of all," death by broadhead is, if not always utterly painless, uncommonly uncruel.

Like all bowhunters worth the title, I practice long and hard and use reliable equipment. I get real close and wait for the perfect shot. Consequently, the animals I shoot generally die within seconds—a fate, dying clean and fast, I can only wish for myself when my turn comes.

WHY DO I BOWHUNT?

Because it's right for me. In fact, at this stage in life, in many ways, it *is* me. I've been bending bows and flinging arrows since I was old enough to kiss the girls and like it—plenty long enough for it to have become integral to who and what I am. Recently I was talking with a man who'd spent years working and adventuring in Alaska. One of the lively tales he told me sheds light (and more) on the discussion here at hand. A few autumns ago, this fellow and two friends (men, I presume, though he didn't say and it hardly matters) were in a remote caribou camp when the weather went unseasonably sweet. Faced with day upon day of cheery sunshine and balmy warmth, and with the summer's bugs all gone to den, the trio found themselves in a de facto subarctic tropical paradise. And so it came to pass, while philosophizing 'round the campfire one night, these three fellows hit on the idea of going hunting next day "butt-nekked" in celebration of the blessed weather and their temporary freedom from the stifling restraints of civilization.

And sure enough, come the halcyon morn, off they went, garbed in boots, hats, daypacks, rifles, and—picture it!—nothing more. "It was fun for a while," the man recalled, chuckling at something I could not know. "But before long . . . well, after the new wore off, it just didn't feel right, carrying a rifle and looking to kill, without being properly attired for the task. Running around naked, we just couldn't take ourselves *seriously* as hunters. The *mood* was wrong, and all three of us felt it."

Well . . . that's me with a rifle. When I'm swathed like a pumpkin in "daylight orange"—the legally required uniform for rifle big-game hunters in most states, for safety's sake—I feel neon-conspicuous, inconsolably incongruous, visibly vulnerable . . . like lurching into the swankest eatery in New York in greasy overalls and a dented hard hat . . . prancing into a Bakersfield biker's bar in Sonja Henie's tutu . . . or strolling around Alaska like Godiva with a gun.

Similarly, when I'm packing a punch that can reach out and kill from a vastly impersonal distance—I just don't feel like I'm *hunting*. Looking to make meat, you bet. And sometimes, in years when the last big game season is running out and the freezer remains empty-bellied, I can justify such slam-dunk efficiency. But it's always anticlimactic: never the hard-core, breath-holding, intensely personal sort of hunting—nature hunting—that I love and crave and always learn from.

For me, packing a high-powered rifle queers the whole hunting deal. With the mood thus skewed, I get sloppy: can't seem to slow and quiet myself enough when moving, or to sit long and still enough on stand. An outsider; an interloper; a misfit—that's Crazy Dave with a rifle.

In contrast, when weaseling through the woods in camouflage, carrying a weapon that requires me to get within slingshot range of my prey, the calming down, the slowing down, the personifying of Ortega's "alert man" all come easy; all come, that is, natural. When hunting well—"playin' in the groove," as a musician friend would phrase it—I feel invisible, confident, alive . . . like a proper predator with purpose and a plan; an actor rather than an audience. And I can hit that groove only with a stickbow and camouflage.

———

As REVEALED by Responsive Management's research, many (though apparently not most) who bowhunt do so to avail themselves of its utilitarian advantages. In addition to the freedom to wear camo, the practical benefits of bowhunting include milder weather during the early archery seasons and the generous length of those seasons . . . the ability to hunt in the breeding season, when elk, turkeys, and other wild things are more active, vocal, and engaging . . . the featherweight weaponry (my lovely custom recurve, with a quiver of five cedar arrows attached, weighs only four pounds, as opposed to twice that for a fully equipped compound bow and nearly triple for a scope-sighted rifle) . . . the secretive near-silence of archery in action, leaving the woods, and the game, undisturbed . . . and more.

Yet, all considered, the greatest attraction of bowhunting has naught to do with practicality or advantage and everything to do with that fine feathered thing called aesthetics—the appreciation of beauty—as in the ergonomic elegance of a serpentine hardwood handle, the patterned grain of natural wood, the sensual undulations of a graceful limb, the second-skin feel and good leathery smell of a well-worn shooting glove, the candy-striped weave of a Flemish-twist bowstring, the Grandma's cedar-chest aroma of a Port Orford shaft, the sibilant sizzle of turkey-feather fletching in flight . . .

And all of it magnified by the bone-chilling bugling and primordial stench of a hot rutting bull—close, upwind, headed your way—on a cool September eve with the golden-apple smell of a quaking aspen grove spicing a clean mountain breeze.

On a survey form, you can't answer questions not asked. Yet, from personal involvement, I can and do read such unwritten and esoteric motives as "aesthetics" and "spiritual experience" into positive answers in the categories of "challenge," "change of pace," "quiet and peaceful," "fun," and, most likely, "other."

———

"MORE," I've learned, almost never means better. And in the hunting woods, it can mean an end to everything good—in some crowded eastern areas, in fact, it already has. Hunting will never be "saved," much less "improved," by more of anything—but only by *better* in everything.

In fact, a primary reason I bowhunt—both aesthetic and practical—is that not too many others do. And frankly, I'd hate to see that change. So please ignore all my romanticizing of the hunt, especially the bowhunt. Football and beer are better, really.

———

YOU MAY HAVE HEARD the old (and admittedly arrogant) archer's cliché: "Where a rifle hunt ends, a bowhunt just begins." As with most generalities, in general it's true. The question thus becomes: "So what?"

In my case, a lot. By haunting the woods with a low-tech weapon, I feel I'm playing a morally informed, ecologically sound, biologically beneficial role in the natural scheme of things. It was, after all, predators—human and otherwise—who sculpted the poetically perfected defenses of the magical beasts we honor as "game" animals . . . even as the game sculpted us. As Charles Darwin discovered a long time ago, and Paul Shepard elaborates so stirringly in *The Tender Carnivore and the Sacred Game,* predator and prey share an essentially symbiotic relationship—an interdependent mutual-aid society. Beasts that are born with or somehow acquire traits that provide situationally adaptive advantages—no matter how minor—tend to live longer and breed more successfully than their lesser competitors. Similarly, those with counterproductive mutations die younger, thus leaving fewer offspring.

After this fashion, genetic "wrong turns" and "failed experiments" are incrementally bred out of a species while the genes of the fittest (best adapted) are refined and reinforced. Via this upward-spiraling cycle, down through the millennia, prey have become ever more alert and cautious, ever harder to catch. In balance, predators have grown ever more flexible, ever more clever, and, among communal hunters such as wolves and humans, increasingly communicative, cooperative, and scheming.

So goes natural selection: an eternal dance of failure and success, death and life, extinction and adaptive evolution. It's not always pretty, but it does always work. Thus did humans and other big-brained, front-eyed, meat-loving predators come to be so deadly bright . . . and our prey so exquisitely elusive—sometimes to the point of seeming illusive.

Through incomprehensible ages spent treading this tricky trail, with predator and prey incrementally upping the ante one on the other—carving, shaping, and honing with the tooth-and-claw scalpels of predation and evasion—did nature patiently sculpt those sublime self-defense organisms we know as the deer. With the arguable exception of the pronghorn, there's no keener eye, ear, nose, and nerve package extant than genus *Cervidae.* And only by continuing the bloody process that built it— predation and evasion—can this delicate pinnacle of wildness be sustained.

Were animal rights advocates ever to realize their fantasies of a world without predation, elk and deer and all the others would quickly lose the razor's-edge wildness that informs, enables, and defines them. In place of continued niche adaptation (evolution), we'd see rapid physical and intellectual decline—*de*volution—even as we see happening today in suburban situations where deer (and lately moose and elk) are becoming half-tamed or "habituated" to people . . . a lurking disaster not only for the animals but for their human neighbors and wildlife managers as well.

As Montana wildlife researchers Michael Thompson and Robert Henderson warn in a professional paper titled "Elk Habituation as a Credibility Challenge for Wildlife Professionals":

> As numbers of live [suburban] elk increased, so would numbers of dead and dying elk; deaths would occur in public view as well as remote habitats. The incidence of diseases passed among elk, domestic animals, and humans might become a matter of public health concern. . . . The eventual result might be lost public interest in conserving and perpetuating elk in their natural habitats. . . . Elk with behaviors valued by society [wild elk] will be increasingly impacted by human encroachment, while "mild wild" elk [habituated animals] will become increasingly common.

Clearly, elk are better off hunted than housebroke, which kills the elkness in them, reducing magical wild things to neighborhood pests.

Rust never sleeps. And since we've either exterminated or severely contained the *Cervidae*'s primary nonhuman predators—wolves, cougars, grizzly bears—and so long as we as a culture remain unwilling to bring back these competing predators across all the deer's range, the task of evolutionary rust removal falls to you, if you so choose, and to me.

And I do it best with a bow.

By taking the role of a *close-quarters* predator, after the fashion of the four-leggeds, I know I'm helping to keep the wapiti wild, win or lose. While today's spectacularly efficient scope-sighted rifles can kill from hundreds of yards removed, such distance-detached elimination of individual animals by invisible snipers adds nothing to the prey's collective education, and little if anything to the shooter's.

Rare is the compound archer who can hit consistent bullseyes much beyond forty yards; for traditionalists, half that is typically tops. The first requirement and ultimate goal of bowhunting, therefore, is *getting close* . . . exactly as must lion, wolf, bear, and Paleolithic human hunters. And to do that, the bowhunter must overcome the prey's seemingly preternatural defenses. Since it furthers the topic at hand, allow me to explain.

FOR OPENERS, cervids see differently than we do.

Our predator's eyes are recessed into the front of our plate-flat faces, spaced three inches apart (on center, on average). Such an arrangement allows the eyes to view an object from slightly different angles, providing predator-essential binocular vision. Whether you're a cougar looking to spring out from ambush at a passing deer, an archer aiming "instinctively" (without sights) at an elk, or a shotgunner leading a sky-rocketing grouse, hunting accuracy and success depend on depth perception. Binocular vision turns that trick. The payback—our limited peripheral vision—matters little to the predatory likes of us.

It matters a lot to the prey, however, who live every moment acutely concerned with the lurking likelihood of predators sneaking in from behind. In response, of necessity, cervid eyes, like those of wild turkeys and pronghorns, have taken an evolutionary hike to opposite sides of the face, where they protrude like little round periscopes, providing panoramic peripheral vision: the legendary "eyes in the back of the head" that so often prove so frustrating to hunters.

Side-set eyes, of course, preclude binocular vision, robbing cervids of keen depth perception. Not that it matters much to them. For a deer, the ability to judge the precise distance to a threat isn't nearly so important as spotting that threat in the first place, identifying it, and tracking its every move.

In order to defeat the deer's visionary eyesight and get killing close, experienced bowhunters disguise their human form with camouflage, employ natural cover, and keep their movements minimal, slow, and smooth: embracing chilly shadows, eschewing always the warm and friendly sun.

Yet, sharp as they are, a deer's eyes provide but a third of its sense/defense strategy. Cervid ears are huge, swivel-mounted, stereoscopic receivers that miss little in the audible world, high frequency to low. Moreover, just as binocular vision grants visual depth perception to predators, deer ears act as auditory range-finders. Make any little sound within a wapiti's extreme range of hearing—a muffled cough, a whispered word, the crunch of a limb, an unfiltered fart—and those big antennae ears will catch it, relaying bearing, range, and threat. Natural sounds, no matter how close or loud, bother wildlife little. But any unnatural noise—especially those that can be identified as human—will win their attention fast, sending them on their way far away. To outwit a cervid's ears and have any hope of getting killing close, bowhunters must move as silently as thought . . . or, more often better, stay patiently put and let the deer do the walking and the noisemaking.

But the deer's remarkable hearing and sight notwithstanding, the nose knows best. The superiority of a cervid's sense of smell to yours or mine is incomprehensible. Not only do downwind deer know when you're about: one good whiff and they have a lock on your position, your range—even, I believe, your intent. (Ever notice how wildlife seem more visible and relaxed when you're just out walking; more invisible and flighty when you're hunting or otherwise seeking them? I suspect this is a true sixth sense derived from the evolved ability of prey species to read the body language of their coevolved predators—including, perhaps, subtle chemical clues to mood transmitted through scent. Regarding this hunch, science has yet to prove me wrong.)

In order to disarm the wapiti's wary nose, veteran nimrods know always to hunt into the wind. Yet for in-your-face bowhunting, that's rarely enough. Having learned through hard experience that a mountain breeze is little more reliable than a politician's promise, savvy archers go all-out to minimize human odor (including, of course, such associative smells as soap, bug dope, and cigar smoke). If you can suppress your scent sufficiently, even if a deer catches your whiff it may overestimate your range and misjudge you—based on the low level of your stench—to be at a marginally threatening two hundred yards when in fact you're lurking at a lethal twenty.

Such finicky attention to detail is rarely required when rifle hunting. When twenty yards or two hundred matters nary a whit. While gunners take pride in their long-range marksmanship, bowhunters (traditionalists especially) are more likely to boast of the intimacy of their interactions with wildlife. (To wit: My average arrow shot on elk is fifteen yards, and I once killed a caribou from just five paces.)

In sum, while killing with a rifle or high-tech compound bow demands only a well-tuned instrument, a modicum of practice, and steady hands, making meat with unaccoutered sticks and string requires a whole panoply of exacting skills, including pinpoint arrow accuracy (necessitating regular practice), wide-ranging woodcraft, physical and mental fitness, and a passion for your prey extending far beyond just hunting it; all the best bowhunters I know are avid amateur naturalists as well: nature hunters. It comes with the territory.

————

ALL SUCH PRAGMATICS and aesthetics aside, I hunt—with bow, rifle, shotgun—to reaffirm natural reality in a made world gone mad. I hunt to provide myself, my wife, and a few good friends with the gift of untainted wild flesh, won hard, with hands, head, and heart, from feral field or forest. I hunt to reconnect, physically and spiritually, to the timeless life-and-death drama that shaped the human body, the human mind, and the human needs for challenge, adventure, and passion.

And I *bowhunt* because—being process-oriented rather than product-driven—it encourages me to enjoy every moment of every hunt, whether I pack out meat or not. For me, traditional archery facilitates hunting satisfaction, and true satisfaction is the best measure of success in any personal endeavor. To take shortcuts to the finish line is to rob yourself of the best that hunting has to offer. In real hunting, a microcosm of life itself, the trip *is* the destination.

I bowhunt—to borrow from Jim Harrison—for all the good it does my soul.

13

Hydraulic Hunting: Wading the Slippery Waters of "No Kill" Angling

> My purpose is not to fight [the fish] but to join them, to watch their tail-waggling riseforms at the end of my cast or their raindrop dimples as I fish down to them in the quiet coves, to feel them against a wisp of graphite, to cradle them in the coolness of their river while it piles against my ribs.
>
> —Ted Williams

No DISCUSSION OF THE ETHICS and practice of modern hunting would be complete without a consideration of a parallel endeavor, which I've come to think of as "hydraulic hunting." Ironically, if not hypocritically, "recreational" angling is enjoyed worldwide by millions of people who would never hunt animals, and many of whom vigorously condemn hunting.

Yet fishing *is* hunting—ask any osprey, otter, or bear.

Yet again, catch-and-release fishing, immensely popular today among fly fishers—differs from all other types of hunting in that no kill is required to consummate the activity: The fish is hooked, played, landed, handled, admired—and released.

Thus, primary among the many intricately interwoven questions to be explored here are:

Why do we fish—including the many among us who disdain hunting?

How, if at all, does "sport" fishing differ morally from "sport" hunting?

And regards catch-and-release angling, what effect does our fishing fun have on the welfare of the creatures we catch, caress, and release?

The answers lie waiting in cold mountain water.

━━━━

I GREW UP ANGLING FOR PERCH, bass, and catfish with cane pole, bobber, worms, and various homemade "stinkbait" concoctions. And still today, my favorite freshwater "eating fish" is the sleek channel cat. Yet catfish prefer warmer waters, and here in the Rockies most water is cold. For this reason, though not this alone, my fishing these days is limited mostly to fly fishing for trout—which, lovely and spirited as they are, I consider second-rate eating fish. But no great matter, since catch-and-release fishing is all about challenge, skill, and being outdoors, and nothing at all about meat on the plate . . . which, ironically, is precisely what opens this superficially benign pastime to increasing criticism and keeps me thinking about why I do it, even as I do it. Like right this moment.

As I mentally scribble these haphazard thoughts, I'm standing prudently shy of crotch-deep in a Rocky Mountain spring lake, having walked a ways to get here. In return for that pleasant effort, I'm assured of privacy, scenery, and unspoiled fishing.

━━━━

MIDSUMMER. Early evening. The only sounds are a light breeze rattling through cottonwoods, the happy twittering of bountiful birds—melodic robins, raucous jays (three local varieties: Steller's, scrub, and piñon), the piercing cries of one killdeer, the signature *peent-peent* of early-shift nighthawks swooping and diving for mosquitoes, the quiescent quacking of a pair of greenheads lurking behind a screen of cattails, and the ratchety chatter of my reel as I strip out line, and the soft wind-whisper of that line arcing behind me, then shooting forward to unfurl and drop, quiet as thought, on the water's mirrored face.

Far above, hard-edged platinum clouds laze through a sky the color of the South China Sea. Nearer, a great blue heron, like some prehistoric memory, approaches for a landing—then spots me and flares away. Sorry, heron. But I'll stay only the evening, and I'll leave the fish all for you.

Lacking breeze or current to invest my imitation mayfly with lifelike movement, I twitch my rod to animate the lure, let it sit a moment, twitch, sit, then repeat the whole artful cycle. Rod hand lifts briskly as line-control hand drops to haul in an arm-length of line, thus shortening and speeding the backcast; a heartbeat's pause as the flying line straightens out behind, then rod shoots forward, whipping the line far out over the water, full speed ahead. Just as the line is about to come taut, I release the length of slack held in my line hand, relaxing the final moment of the cast so that weightless fly, hair-thin leader, and weighted line drop gently and quietly to the water . . . one, two, three.

Form informs fly casting. And that form, along with knowledge—of trout, the insects they eat, and the quirky personalities of mountain water, moving and still—define fly fishing. It's a song of the senses; a celebration of the spirit.

Fly fishing is to bait fishing as ballet is to bowling, and to spin fishing as chess is to checkers. The gear selections alone are daunting, with fly lines that float, sink, shoot, defeat wind or ice, and come in a variety of weights; rods of many materials, lengths, flexure, and weights; plus reels, creels, nets, leaders, tippets (leader-leaders), waders, wading shoes, and a brain-boggling multitude more.

And of course the flies: the artful heart of the matter—minuscule flower arrangements of feather and fluff, more often than not fashioned from natural materials, bearing names variously elegant, comedic, and bawdy: Royal Coachman, Parachute Damsel, Blonde Humpy, Green Butt, Woolly Bugger, Madam X, Gold Bead Prince, Bitch Creek Nymph, and hundreds more.

You could devote your kid's entire college fund to acquiring fly fishing *stuff*—and never exhaust the options. Hence the snob factor so commonly associated with the sport.

But it doesn't have to be that way. You don't have to be a dentist or investment banker, wear tweedy threads, or cast like Brad Pitt in *A River Runs Through It* to enjoy fly fishing. Outfitted with an inexpensive rod, a simple reel spooled with 5-weight double-tapered floating line, and two each of half a dozen basic fly patterns—plus a few essential skills learned from books, magazines, videos, friends, and practice—you're set.

I WAS FIRST EXPOSED to the magic of fly fishing at age eight while camping with my folks on some lovely little Rocky Mountain stream whose name is long forgotten. My parents weren't fishers, and this was my first encounter with trout. But I had worms in a coffee can and the fishing was good, so I caught a few in spite of myself.

Yet the focus of this recollection is not me but the "old guy" camped nearby. He was a fly fisher, the first I'd ever seen in action, and he caught a lot. He didn't have much to say, but seemed willing enough to let me shadow him and watch . . .

. . . Watch as he surveyed the air and water to identify the specific aquatic insects hatching then and there. Watch as he selected the appropriate imitation from dozens hooked into the wide wool band circling his fedora. Watch as he studied the sun-spangled water, mentally mapping likely trout haunts. Watch as he ticked his willowy wand of Tonkin Gulf bamboo to and fro, steady as a metronome, the fly never touching the water in front or the vegetation behind, "dry casting," letting out a bit more line with every repetition, until the rod tip pointed and stopped and the line shot forward and fell, just so, daintily depositing the tiny lure exactly where he wanted it.

And often as not, exactly where the trout wanted it as well. A quick silver flash, amorphous yet distinctive, as the fish rose and nosed the lure; the old guy's rod rising quick in response. After playing the fish until it grew tired and became compliant, my hero would finesse it in, net it, admire briefly its writhing form (and grant me a glance), then grimace as he smacked its head against a rock, deftly gutted it, and slipped the colorful corpse into a wicker creel lined with cool wet grass.

This, of course, was back in the days when catch-and-devour was what was—years yet away from the time that dams, irrigation, overgrazing, pollution, human overpopulation, and trend-setting movies would conspire to mandate a new fly fishing paradigm: catch and release.

But even if he'd never killed or even caught a fish—you could see this in his every move and gesture—the old guy would still have been in heaven . . . standing heart-deep in that perky mountain stream, the icy press of recent snowmelt rushing all around, immersed waist-deep in it all, writing love letters to life with an eight-foot quill of Vietnamese bamboo.

Back home, the closest I could come to imitating the old guy's grace was flipping little cork "popper bugs" at cow-pond perch with a cane pole. I caught a few that way, some bullfrogs as well, and ate them all (well, only the fat aft legs of the latter). It was fun. And in that place and time, for me, fun was a revelation, found only in the out-of-doors.

Time passed, and with the help of increasingly frequent outdoor escapes—hiking, camping, hunting, fishing—I survived my horrid high-school years without getting killed and only occasionally caught.

In 1968, aged twenty-two, I ran away to the Marines. In 1974, I found myself beached in southern California, where the Corps had deposited my remains. (Actually, I resigned.) There I inhaled deeply, meditated on ocean waves, chased surfer girls and hippie chicks, and caught a precious few. And all of it was good. Yet something essential was missing from that narcissistic SoCal scene.

Thus, searching for that something, did it become my wont, once a month or so, to beg out of work (I edited a motorcycle magazine) at noon on a Friday and fly my old VW bus for all she was worth (fifty-two with a tailwind) out across the hellish Mojave Desert (land of broken glass and shattered dreams) to the cool green relief of Sequoia National Forest. There, I'd follow some lonesome dirt road to its end, sleep for what was left of the night, rise with the birds and backpack several hours up some gurgling mountain stream, make a spartan camp, fly fish until dark, eat my catch for dinner or go without, sleep until daylight, hike out, drive home, and show up at work happily exhausted and more or less on time Monday.

Friends called it fanaticism. I called it necessity.

I can't say exactly when and why I began practicing what *Audubon's* ever-clever Ted Williams calls "the voluntary surrender of victuals." But I do know it came only after, and had to do with, eating enough trout to gag a grizzly bear.

———

NOW I AM OLDER than the "old guy" who first fanned my flame for fly fish-ing—wayworn and a little drifty as I stand here, crotch-deep in pure Rocky Mountain spring-lake water, reciting the fly fisher's mantra: cast, pause, twitch, pause, twitch, pause, cast. The sun slumps toward the distant Pacific, my home of long ago, as the late-evening birdsong hits a crescendo. The fishing is slow, but so am I; it's a fit.

IT'S CHALLENGING, it's artful, it's fun and relaxing, this catch-and-release angling. And on its glossy surface, at least, it seems a right gentle way to interact with wild nature. Yet what might be the *trout*'s take on this? What effect might having a hook set in *your* mouth, being forced to struggle—for your very life, so far as you know—at the long end of a line, then hauled, exhausted, from the element that sustains you, only to pant and gasp while being unhooked, hoisted aloft—a living trophy—and photographed, then finally released . . . how might such an experience affect *your* day?

We can't compare fish and people. Yet, sniffing cautiously along this tricky trail of thought, a growing number of concerned fly fishers are turning to barbless hooks, which mandate a bit more mastery to land a fish but also limit tissue damage while easing and speeding release. And for a fine-tuned few, even this is not enough.

I offer for example Stanton Englehart: art professor emeritus, Southwest landscape painter, and fly fisher supreme. Englehart's concern for the trout he and his wife Pat catch and release led him some years ago to invent "pointless" fly fishing. "Frequently," Englehart recalls, "after landing and releasing several fish, Pat would turn to me and say, 'This is pointless.' And it was: Though fly fishing remained challenging, handling trout had become redundant and even distasteful. So one day I acted on Pat's subliminal suggestion and clipped the point off my hook. I found that I could still enjoy the pleasures of reading the water, selecting the perfect imitation, casting, presentation, and 'touching' the fish with my fly, all without harming it or even moving it off its hold."

Which prompted me to challenge Stanton: If you're all that concerned about the welfare and comfort of fish—why not quit fishing? Stanton was ready for me. "The rhythm and mystery of the river are magic to us," he explained, thoughtful eyes twinkling, "and that long fluid line is the thread that connects us to it all. We could never give it up."

Even so, Englehart readily acknowledges that "pointless" fishing isn't for everyone. Catch-and-release does have its place. "As much as I hate to see fish harassed and possibly harmed by excessive handling," he admits, "catch-and-release waters are vastly preferable to the low quality of put-and-take fisheries, where every fish caught is killed."

This is true: Catch-and-release angling (as opposed to catch-and-kill) leads to more and bigger fish, more and bigger pleasures for serious fishers, as well as better-balanced aquatic and riparian ecosystems. As legendary angler and angling writer Lee Wulff (the "father of catch-and-release fishing") observed decades ago: "Gamefish are too valuable to be caught only once." "Too valuable," that is, ecologically and aesthetically as well as recreationally. Today, Trout Unlimited continues to champion Wulff's conservation wisdom through its credo of "limiting your kill instead of killing your limit."

Of course there are other views. One of them is voiced by Stephen Bodio, among America's most literate sporting writers. Though primarily a catch-and-release man himself, Steve advocates the occasional killing and eating of trout as a means of reminding ourselves of the ancient visceral connection between human and fish, predator and prey, that gave ancient birth to today's "blood sport."

In other words: Killing and eating even the occasional fish reminds us that fishing is—in fact and instinct, if not always effect—hydraulic hunting.

Nor is Bodio alone in finding it easier to stomach the killing and eating of wild creatures than using them as playthings. This view, I suppose, could be tagged catch-and-*use,* as opposed to catch-and-*abuse.*

Indirectly endorsing Bodio is Ted Williams, who boldly expresses equal passions for preserving and viscerally touching animate nature. "My favorite tip for catch and release," Ted jokes, "comes from the Newfoundland Provincial Government: 'Do not use gaff to handle fish.'" Railing against the management problems caused by the "holier-than-thou, no-kill types who prate and simper and misquote Shelley" while piously refusing ever to kill a fish under *any* circumstances, Ted cites the ongoing dilemma with brook trout in Rocky Mountain National Park, Colorado:

> Biologists there are trying to eliminate non-native brook trout to aid the comeback of threatened native greenback cutthroats. Releasing a fish that needs to be removed from the ecosystem is worse than Zorba the Greek's unpardonable sin of not going to a woman's bed when called. But the fly fishing snobs insist on releasing *all* trout. One manager told me that 80 percent of the brookies there have been caught and released, in spite of the park's best efforts to educate anglers to the contrary.

Montana has fared somewhat better. There, in the Bighorn River a few years ago, a congestion of adolescent brown trout were suppressing the rainbow population, preventing both rainbows and browns from reaching full growth. As an antidote, managers asked anglers to keep and eat a few mid-sized brownies each time they fished the Bighorn, while releasing all rainbows and big browns. Fishers complied, and the Bighorn is now a happier, healthier place . . . for all.

Which reminds me (never mind how) of my great friend and near neighbor, George Hassan, who has found a happy, if somewhat eccentric, compromise that allows him to maintain the aesthetic purity of catch-and-release while occasionally enjoying the sizzling pleasures of catch-and-gobble. When fishing for the pleasure of it (virtually every day of his life), George is a no-kill fly-master. But on those rare occasions when he gets orders from Nancy to "get us some fish for dinner," George uses worms. Thus does one man keep "the quiet sport" quiet.

"The point," says George, "is knowing what you're after—fun or food—and pursuing it accordingly. Worms have caught more fish than all fly patterns combined. But there's hardly any challenge to it; nothing much to learn. And since trout tend to swallow live bait, hooks can't safely be removed; worm fishing is for keeps. By contrast, trout usually take flies in the corner of the mouth, and can be released without noticeable injury."

Without *noticeable* injury, yes. But what of unnoticed injury? What about pain? Do hooked and handled fish suffer, physically and/or emotionally? And if so, how much? Tough questions, these, with no pat answers. Scientists generally admit as much, though often expressing a cautious "no brain, no pain" bias. Not so People for the Ethical Treatment of Animals—best known by their acronym, PETA—who know it all.

In a letter to Walden Pond State Reservation in Massachusetts, PETA spokesman Davey Shepherd requested an end to all fishing in those hallowed Thoreauvian waters. Why? Because "fish have individual personalities . . . they talk to each other, form bonds, and sometimes grieve when their companions die." These all-knowing folk also condemn keeping pets and riding horses. But PETA's brightest moment (to date) was when it petitioned the town of Fishkill, New York, to change its name to Fishsave—either unaware or uncaring that in that neck of the eastern woods, in that usage, "kill" (derived from the Dutch *kil*) signifies a mountain creek or channel, not death.

WELL, THEN: Where *do* we stand on the question of human recreational values versus wildlife welfare, in this case fish?

All things considered, and admitting a personal bias, it strikes me that a swiftly landed, gently and promptly released, barbless-fly-caught fish likely experiences far less pain, physical or emotional, than your average, say, backpacker or bicyclist or jogger endures for hours or days on end, calling it fun. Obviously, joggers et cetera *choose* to endure their pain, while a hooked fish has no choice. But my point here, insignificant as it may be, is not electivity, but relativity.

Sadly, there are exceptions to the "swiftly, gently, promptly" priorities of conscionable catch-and-release. In high-profile ("blue ribbon") trout and salmon waters worldwide, hook scarring and other cumulative catch-and-release injuries are all too common. And a percentage of hooked and handled fish die from the trauma.

Yet people always have fished, always will. In North America alone—according to the U.S. Fish & Wildlife Service's most recent (1996) figures—some thirty-five million women, children, and men angle every year. And of that lot, an impressive twenty million claim to be full-time or part-time catch-and-releasers. Somewhat more precisely, the *Wall Street Journal* reports that in 1997 Americans purchased 29.4 million fishing licenses. Add to that the children, retired people, and other license-exempt categories in order to approximate the Fish & Wildlife total.

Even from the fish's point of view, this is not so bad. Through license sales, special taxes on sporting equipment, and outright donations via conservation-oriented "sporting" groups such as Trout Unlimited (whose hundred thousand members raised $10.7 million for fish and fisheries conservation in 1998), the Izaak Walton League, and others, America's anglers contribute a king's fortune annually to help purify polluted waters, restore ruined riparian areas, purchase critically threatened aquatic habitat, and otherwise restore and preserve robust fisheries—all of which works to the benefit of entire ecosystems (otters, ospreys, bears, and more), as well as for fish and fishers.

Hunters do the same. The Rocky Mountain Elk Foundation, citing the example I know best, at the close of 1999, notwithstanding the rapid growth of its membership and infrastructure in recent years, still managed to direct a whopping 83 percent of its income into such worthwhile "on the ground" projects as rescuing critical wildlife habitat from development through direct purchase, wildlife habitat improvement, and education.

PETA, meanwhile, recently purchased a new $2 million headquarters building.

———

BUT ALL ELSE ASIDE, bottom line, any way you cut the bait—pointless, barbless, catch-and-release, catch-and-clobber—the onus is on ethical anglers to do what we can to minimize the abuse of fish and their habitat . . . in precise parallel to the responsibilities of all ethical hunters.

As I see it, the keys to fishing with a clean conscience are compassion, moderation, voluntary cooperation with sensible management goals, and good old-fashioned common sense. To wit:

- If catch-and-chew is your goal, pursue it in put-and-take (hatchery stocked) waters or where it will improve the aquatic neighborhood—helping, say, to cull those domineering Rocky Park brookie aliens that Ted Williams decries, or combating the voraciously predatory, nonnative lake trout threatening to decimate the native Yellowstone cutthroat population, thus eroding the entire Yellowstone ecosystem.
- In waters supporting self-sustaining trout populations, where fishing pressure is heavy, where fish densities are low, or where wild native populations are struggling for survival, practice barbless catch-and-release.
- In intensely fished areas where trout are handled often enough to show battle scars—refuse to participate and encourage others to refuse likewise.

I'd like to say I'm not proselytizing for fly fishing—that I don't give a rat's rump whether you do or don't. But it's just not so. Because I believe there's a good chance that anyone who gives fly fishing a fair chance will come to love it as I do. And to love fly fishing is to love those sparkling stretches of water that "your" trout call home, along with the scenic landscapes those lakes bejewel and those rivers run through. And you, I, we, will fight a lot harder to preserve what we love—enlightened self-interest at its natural best.

———

MEANWHILE, back lakeside, another good day is almost done for. My naked legs have long since gone blue and goosey from prolonged exposure to icy water. A warming campfire beckons.

But not quite yet. A twilight caddisfly hatch has just erupted, and this lovely little lake's formerly slothful fish suddenly are rising everywhere to feed, some leaping high for their dinner (and, we may fairly imagine, for joy), slapping the water like beaver tails. Others—big, cautious browns, most likely—expose only their lips to daintily, though quite audibly, suck tiny emerging insects from the filmy surface. No catch-and-release going on here.

Of course (Murphy's stinking law), it's rapidly growing too dark to fish. Acquiescing, I execute what I grudgingly decide will be my penultimate cast, randomly selecting a nearby rise-ring as a bullseye . . . and—*Holy Molé!*—the water literally explodes as my #14 Elkhair Caddis is snatched and hauled hungrily under.

With the speed of long-practiced reaction, I lift the rod sharply, straightening the line and setting the tiny hook. This prompts Moby Trout to sound and run, deep and long, ripping out line as my reel screams with excitement.

Where moments ago I was languid and lost in thought, now I'm hopping with adrenaline and my heart is sledgehammering inside my chest. By the gods, *we're alive,* this fish and I!

After thirty yards, the colorful creature abruptly ends its run and just lays there, stolid as a boat anchor. I slack the line to let the fish rest a moment, then apply slow pressure with the reel, regaining a few precious yards. Predictably, this prompts another run, only slightly less ferocious than the first.

And so goes the contest—line out, line in—as I attempt to finesse the fish to the finish line, establishing the best compromise I can between getting it over with fast (for the trout's sake) and not getting so heavy-handed as to snap the tenuous tippet (for both of our sakes). My rod arm is aching and my knees are shaking, the latter no longer from cold.

In good time, at last, my worthy opponent tires and begins to come. I work "him" gradually closer, getting a first good look in the evening gloom: a cuttbow, or rainbow/cutthroat hybrid, crimson-sided from cheeks to tail, multihued and luminescent—the most beautiful trout of them all, I say. And as big as . . . well, we'll pretend that doesn't matter.

Rushing to the release, I tuck my rod under one arm, reach carefully into the water (no net and none needed) and slide both palms beneath the temporarily docile animal. While one hand gently encircles the tail, the other eases forward to let slip the hook. The fish makes no complaint. That done, I support the brilliant creature upright in open cupped hands, helping it to rest and recover.

Seconds pass, then, with one strong torque of tail, the gorgeous salmonid flashes away, diving for the darkened depths.

———

WHY DO WE FISH—including the many among us who will never hunt? The answers lie waiting in cold mountain water.

PART III

Onward, Through the Fog

Just as game animals are the truest indicators of quality natural environment, so hunting is the truest indicator of quality natural freedom.

—John Madson

14

Gender Be Damned: A Woman's View of (Men and) Hunting

> She is a lady of the wild things, including the instinctual wildness within herself.
>
> —Christine Downing

A FEW YEARS AGO, having learned she was a hunter, I emerged from my social carapace to befriend a "neighbor lady" whose cabin shares the mountain with ours. That lady, who has since become family to Caroline and me, is Erica Fresquez, as passionate an "outdoorsman" as any I know. I'm including Erica's story here because she brings life, personality, and feeling to the otherwise hypothetical "woman hunter" I discuss in the following chapter—and because she's fun to read. This chapter, therefore, is Erica's story.

THE NEW DAY is a streak of gold on the eastern horizon, the woods eerily silent, when I find my place beneath tall timber on a frosty north-facing Colorado mountain slope. A stately old ponderosa giant will be my easy chair for the morning, its trunk my backrest, its two huge exposed roots, like muscular arms, reaching out to cradle me. Though not many women would want to trade places with me just now—alone in the predawn dark, deep in the woods, on a cold mid-November morning—I feel a sweet, familiar peace as I settle in to wait.

I am a hunter.

———

MY HUNTING LIFE grew out of a family tradition—or, rather, out of my exclusion from that tradition.

My father, a sheep rancher, was the son of the son of a sheep rancher. Their home, and mine, was the Bar-Guitar Ranch near Picacho, New Mexico, a tiny village tucked away in the Hondo Valley, a place of endless rolling hills creased by the Hondo River and speckled with sheep and apple trees.

The Hondo landscape is a canvas painted with a thousand fond memories of my childhood. With my mother, Lupie, at his side, my father returned from the navy to run the family ranch and raise a family. In 1962 I was born, fifth of six children.

Five years later, while on duty as a reserve sheriff's deputy, Dad was stabbed, almost killed, and left temporarily incapacitated. The incident was doubly heartbreaking in that it forced us to relinquish the ranch and move to nearby Roswell.

Although no longer a rancher, Dad remained a hunter. Venison was the richest part of many meals back then, accompanied by refried beans, fresh green chilis, fried potatoes, and Momma's famous hand-rolled tortillas. With elk replacing venison, it remains my favorite meal to this day.

I always knew when autumn had arrived by the change in seasonal colors—inside our house as well as out. With the approach of deer season came the reappearance of Halloween-orange hunting clothes, scattered around on the furniture. More poignantly, I recall my two brothers' special bond with my father at that magical time of year. Their voices were more alive, their laughter more full and connected.

From my first recognition of the magic of hunting season, I longed to be included. But I was too young, and I knew it. Since hand-me-downs were a necessity in our family, I waited until I could loosely fit into my brothers' cast-off hunting clothes before inviting myself along. My father's answer was no. In his view, hunting camp was no place for a girl.

"I can scramble a mean egg," I countered desperately.

No.

These annual, all-male camps were a tradition that continued until my father died from an aneurysm in 1978, on his forty-fifth birthday. I was only sixteen. I still needed him. We all did.

————

THE MORNING LIGHT brightens above the eastern slope of the little sidehill valley I'm haunting, sifting softly through the trees. Gradually, the forest comes alive: ravens croak, gray jays fuss and scold, pine squirrels ratchet and leap about; a dawn breeze, like a yawn from the sun, whispers through the forest.

The woods and understory here are so thick they limit my vision to less than fifty yards in any direction. Yet I feel confident in my setup: Several game trails intersect nearby, all freshly churned with elk tracks and dotted with their fall pellet droppings. Starting to shiver, I wriggle my fingers and toes in an attempt to generate some heat.

As usual, I've chosen to hunt the last of Colorado's three annual rifle seasons, counting on mid-November snow to move the elk down from the high country and provide good tracking. But this year, rather than snow, we have cold, clear mornings and warm, dry, crunchy afternoons. Yesterday was a good example.

My guide for the day was David. In the recent past, David and I have teamed up to chase not only elk but deer, turkey, and small game. With his elk in the freezer since archery season, David came armed yesterday only with a backpack and an excuse (me) to get out into the wapiti woods again.

Our adventure started well before dawn as we crept quietly into a mixed conifer and aspen forest on a dry, south-facing slope. David had seen "plenty elk" there during September and thought it was worth checking out. As if we were bowhunting, we moved slowly and deliberately, stopping often to look, listen, and sniff the air for scent, trying to be silent and doing a pretty good job of it, at least for a while.

But as the morning warmed and the night's frost thawed and dried, we found ourselves shaking our heads in frustration as crispy leaves and branches crackled underfoot—nature's burglar alarms. Yet with "no sign nowhere," as David put it, sitting on stand was useless. Our only option was to keep moving slowly on, now scouting more than hunting.

We never did find the elk that day. David surmised they had moved to the cooler, moister, north-facing slopes—like, consequently, the one I'm watching now. What I gained from the experience, reinforced by David's advice, was to come here this morning alone, find a good setup, stay put, and "let the elk do the noisy walking."

A Boeing 737 whines high overhead, like some great silver mosquito dragging a long white tail across an otherwise perfect sky. I greet the disturbance with a crooked smile, aware that next week, like last week, I'll be up there in one of those flying tincans myself, paying my way in the world as a flight attendant. But for now, I am hunting.

———

WITH THE EXCEPTION OF ANNIE OAKLEY, my outdoor exemplars and mentors have all been men: a mixed blessing.

Following my father's death, I spent my $800 inheritance on a used car in order to return to the Hondo Valley as often as possible, visiting friends and family. In an attempt to recapture the good old days, I volunteered my help to every rancher I knew. I gathered and marked sheep. I rounded up cattle and branded them. I rode fence for my Uncle Pete, an all-day job on horseback, which I felt honored to be trusted with alone at such a tender age.

But among all those good Hondo folk, there was one special friend. His name was Sam Montoya, sixty-nine at the time. Soon, my father's best friend became my best friend. It was Sam who taught me the gentleness required to saddle a spirited horse. Sam talked of the importance of education, and I learned by his example the joyful value of a warm heart. Sam taught me how to find my way home from the hills, and together we rode those hills. And it was Sam who introduced me to hunting.

One evening at his house, after a long day of mending fence and tending livestock, Sam reached up high on a shelf, pulled down his old binoculars, placed them before me on the kitchen table, and announced: "These are for you, Sweet. Next week, we go deer hunting, you and me!"

That same night, I got a crash course in handling a rifle. When I asked my brother Ralph to show me how to shoot one of Dad's rifles, he chose an old Winchester Model 94, an iron-sighted, lever-action 30-30. Ralph showed me how to center the front sight blade in the rear sight V, cautioning me never to cock the hammer or put my finger on the trigger until I was ready to shoot. After checking the chamber to be sure it was empty, and picking a spot on his bedroom wall, the dry-fire practice began.

Following an excruciating week's wait, deer season finally opened. For Sam and me it would be a working hunt—combining ranching with watching for deer and talking about them. On the third day, as I was riding fence on horseback with Dad's 30-30 across my lap, a muley buck jumped up from his bed—giving me my first tangy taste of hunter's adrenaline. Leaping from my horse and aiming carefully, I took the shot . . . a *running* shot. Instantly, incredibly, the buck went down. (Today I would never consider a running shot, knowing the high risk of wounding rather than killing. But then I was young and dumb, and very lucky.)

Sam, who'd been riding well ahead of me, heard the rifle's crack, spurred his horse, and came galloping back in a cloud of dust, just like a Wild West movie. Sam looked at the deer, then at me, and for a long time said nothing. He just grinned a grin I interpreted as pride. At last he exclaimed, "You did it, Sweet!" grabbed me with both arms and pulled me into a hug.

My 4 × 4 buck was huge in my eyes. And killing him clean, on the run, with just one shot—my first shot ever with a rifle—was very near miraculous. Yet in Sam's creative retelling of the event, I became Annie Oakley reincarnate.

"That buck," Sam boasted far and wide, "was streaking away fast as a mountain lion when Erica made a perfect shot—over her shoulder, twisting in the saddle, aiming through a wall of dust churned up by the pounding hooves of her running horse."

Boy! What a word-picture he painted . . . Sam Montoya, my favorite artist.

EIGHT O'CLOCK and the sun is full up, though here beneath the forest canopy it's still dark and cold. Fighting the urge to stand and stretch, I remain motionless, leaning into my tree, a shivering bundle of heightened senses, hoping the scolding squirrel above me isn't giving me away. Waiting, watching, listening—and when the squirrel finally shuts up, hearing.

But hearing what? A jay hopping around in a bush? A squirrel scurrying through the leaves? Or something bigger, trying to walk quietly? I recall David's advice to weigh every sound, listen to the cadence of its movement, anticipate its approach, be ready. Weigh every sound: This sound is far too heavy for bird or squirrel. Deer? No, still too heavy and brutish rather than dainty. Bears walk softly on padded paws, rarely making any sound at all, and most are in den by now anyhow. That leaves only . . .

Here we go! I struggle to control my breathing, but with little luck. I can feel my heartbeat pulsing through my entire body. The sounds continue: big animal, coming closer. Swimming hard against a flood of adrenaline, I bring my 30-06 to my shoulder, slip my index finger into the trigger guard, flick off the safety, and peek out around the scope sight. With an either-sex tag in my pocket, the season half gone, and my mouth watering for elk steak, I don't intend to be picky.

And here he comes, closing fast from starboard . . . one, two, three, four antler points to the side: a lovely legal bull. Isn't it the way with hunting? You walk and you wait, often for weeks, and then it all happens so *fast*.

As the bull draws closer I center the cross-hairs just behind his left fore-leg—at only twenty yards, he fills the scope. When all is perfect, no chance for heartbreaking error, I squeeze. The rifle roars and pounds my shoulder, its echoes rolling like thunder down the mountain.

Confident of my shot, I wait for the bull to fall. But he doesn't. Rather, he brakes to a sudden stop and looks directly at me. As I bolt another round into the chamber he makes a sharp left turn and starts back up the hill he just came down. Refusing to believe I could have missed, I bring my rifle up for another shot, find the bull in the scope—and watch him collapse. The great beast's stomach swells and falls with one last breath, then all is still. I glance at my watch: 8:06 A.M.

After waiting a full minute, ready to shoot again if necessary, I click on my rifle's safety, open the bolt, place the weapon gently on the ground, stand, and go to the bull.

My beautiful wapiti!

Kneeling at his side, I stroke the rich black hair of his neck mane, oddly soft, like rabbit fur. His nose also is amazingly soft, feeling pleasantly warm against my chilled fingers, as I hear myself saying, *Thank you. Thank you.*

Suddenly my vision blurs and I wonder: What have I done? And why? The answers, as if spoken by somebody else, come right back: If necessity is defined as something you cannot do without, then hunting is for me a necessity, tears and all.

Drying my eyes, I reach for my sheath knife and begin the hard work of field dressing, joyfully accepting my responsibility to my prey, and to myself. More than anything right now, I want to share this moment with David.

I miss my father a lot, especially at times like this, sitting here alone beside this great bull elk. I imagine his smile and how proud he would be— even though he always maintained that hunting was "not for girls." I would love to have hunted with my father, even just once, and can't escape the disappointment that it never happened. Yet I am blessed to have a friend like David, who has given me admission to his very private world, and I hang on gratefully.

The morning is middle-aged by the time the field dressing is finished and I've hiked down the mountain and driven home, hoping to find someone to help me pack out my elk. (For now, he'll keep just fine, eviscerated and lying on pine needles in cool shade.) As luck would have it, David is in town, so Caroline and I leave him a note and are about to go it on our own when we run into Koby, a bowhunting, firefighting neighbor who has the day off. Together, the three of us head back up the mountain to my prize.

Young and strong, Koby could probably pack a whole elk out in one trip. But before we can put that hypothesis to the test, David shows up, having found our note, bearing a big hug of congratulations and his well-used freighter pack. A few minutes later, proudly I lead my helpers down the mountain: Between David, Caroline, Koby, and me, we get the whole huge animal out in just two trips—meat, hide, head, and "horns."

———

SOME FOLKS, women as well as men, have called it fanaticism, this passion of mine for the hunt. Others, echoing my father, have suggested that hunting is "unladylike." I just call it love. And like all real love, it isn't always easy.

When I first moved to Colorado, my challenge was to make new friends who shared my passions for horses, hunting, fishing, and nature. As it turned out, finding people with those common western interests was easy. The hard part was being taken seriously as a woman. It was almost like being a child again and trying to get my father to take me hunting—an untenable "girl" trespassing onto traditional male turf.

My first Colorado elk-hunting mentor was a crusty, middle-aged mountain man. Boyce took one look at me (5'2" and 115 pounds), and through a puff of cigarette smoke predicted: "You won't last through your first winter up here." But looking back on it now, fifteen winters later, I don't recall being offended by his remark. It just made me determined to prove him wrong. And for all his bluster, Boyce soon relented and showed me where and how to hunt elk—his way, at least.

Although I never came off the mountain with an animal while hunting with Boyce—he was always quicker on the draw than me and gave no quarter—I learned many valuable lessons about wilderness survival, elk hunting, game care, and packing meat on horseback down steep mountain trails. Though we had little else in common, elk, horses, and wilderness bonded our friendship.

A little later, I met Tommy, another "she won't last" doomsayer. The first morning that Tommy guided me out to hunt, I took my first wapiti—a spike bull killed clean at a hundred and fifty yards. After that I continued to hunt elk, though with the basics of where and how now under my belt, I went out mostly alone—until I met David.

My good friend Karen Kebler and I had just spent a week hunting deer, unsuccessfully, near her home in Washington state, followed by an equally meatless week chasing elk in Colorado. On Karen's last night here, I got a call from a neighbor who'd just hit a mule deer that had darted in front of her car. She'd tried calling David, knowing his fondness for fresh roadkill, but ironically he was down in New Mexico (near where I was raised), hunting deer. So Chris called me. The fork-horn buck was head-hit, its body undamaged, and Karen and I were thrilled to go out in the midnight rain to retrieve the meat.

A few days later, David, having heard about my roadkill rescue, stopped by to offer congratulations. David's first words to me were: "I'm impressed!" Before the conversation ended, he had offered to introduce me to wild turkey hunting. This would be a whole new adventure for me: sneaking around in camouflage, trying to get shotgun-close, rather than walking or riding around in blaze orange with a long-range rifle. It was a big step up in hunting challenge, and I could hardly wait.

———

APRIL FINALLY ARRIVED, and off David and I went to the turkey woods. By the first hour of that first morning, I was hooked on the wild turkey chase. David, already grooming me for bigger things, said turkey hunting was "great training for bowhunting elk in rut." Better yet, a new friendship was developing, based on mutual respect—gender be damned—and a shared love of all things wild and free.

And as a woman, a new fashion statement was evolving for me. As a rifle hunter, my field wardrobe consists of ugly orange, which I wear because legally I have to. But put me in camo and turn me loose in the woods, and I know I *belong* out there. This duality, between the mandated orange of a rifle big game hunter and the voluntary camo of a turkey (or bow) hunter, strikes me as a metaphor for my double lives as "prim flight attendant" and "gnarly mountain chick" (I've been called both): one is a uniform, the other is *me.* Unlike humans, wild animals care nothing about your sex, weight, whether or not you wear earrings, or the length of your hair. They only complain (by running away) when you make too much noise, silhouette yourself on the skyline, or smell bad.

Being a woman who hunts, though challenging in many ways, especially at the beginning, has proved to be one of the most fulfilling aspects of my life. Today I'm a capable and ethical hunter, fisher, and (as David likes to kid me) "outdoorsman." I have not only boundless enthusiasm for what I do but, perhaps more important, a solid understanding of why I do it.

15

Women and Hunting:
A Man's View

Might [women hunters] be taking back something that
began to be denied us all, roughly ten thousand years ago,
in the shift from hunter-gatherers to agrarian cultures that
spelled the dawn of patriarchy?

—Mary Zeiss Stange

JUDGING FROM THE WOMEN HUNTERS I know best—Erica, Anita, Sandy,
Diane—together with what I've learned from Paul Shepard and others, I
agree absolutely with Professor Stange. I would, however, specify that the
"taking back" is not by political force, but direct action.

Consider my hunting buddy, Erica Fresquez. The same year of the successful elk hunt described in the previous chapter, Erica also bagged a deer
and a turkey. And she did it all alone. And that's exceptional. While solo
hunting may or may not be "superior" to social hunting, it takes more skill,
motivation, and guts, and therefore offers a challenge that most female (and
many male) hunters are hesitant to accept.

Moreover, Erica takes no shortcuts: She has never hired a professional
guide, hunts on foot and mostly on public land, wants no part of baiting,

hound-treed animal executions, or gadgetry, disdains wildlife ranching and canned killing, and in every way personifies the fairest of fair chase.

Thanks to Erica's skilled and persistent hunting and fishing efforts (augmented occasionally by roadkill scavenging), the Fresquez freezer stays fat with what Paul Shepard calls—and Erica knows to be—"the best of foods," all those neatly wrapped bundles of wild meat sandwiched between big bags of New Mexico green chilis, the hotter the better. Dinner at Erica's is truly special.

Yet, as her candid self-telling suggests, Erica's long struggle for acceptance as a serious hunter was never easy, and still isn't today. In this and other ways, Erica's story is Everywoman hunter's story. It includes such commonalities as her late initiation (by sixteen, most boys with hunting elders already have years of experience)—resulting from the common male conviction that hunting is a "guy thing"—compounded by the disapproval, subtle or otherwise, of nonhunting women.

Commenting on the irony inherent to this cultural double-squeeze—both men and women are uncomfortable with the concept of women hunting—Mary Zeiss Stange writes in *Woman the Hunter*:

> To the extent that hunting has served both patriarchy and feminism as a root metaphor for men's activity in the world, Woman the Hunter is a necessarily disruptive figure. She upsets the equilibrium of the conventional interpretations on both sides. . . . In the radical feminist vision, the mere involvement of any woman in such male-identified activities as hunting and shooting is enough to qualify her as a creature of the patriarchal system. . . . Yet—and here is a crucial irony that calls into question all such facile bifurcations of social reality—women who assert themselves as equal in skill and power to men, who take men's equipment into their own hands for their own use, are perceived by patriarchy's "boy-men" as intensely troubling, divergent and threatening . . . indeed, as feminists!

But neither side need worry. As Stange contends and I concur, women who hunt, including my friend Erica and Dr. Stange herself, "do not appear to be hapless pawns of patriarchy. They do not seem to want to be, or to act like, men."

Yet most women must depend on males as hunting mentors. In fact, most women hunters I've talked with say they weren't introduced to hunting, hadn't even considered it, until a man in their life—father, husband, friend—encouraged their company afield. This suggests that only a minority of women hunters shared Erica's experience of being actively denied as a child, while the majority had other interests and never thought to ask.

———

I AM, I FLATTER MYSELF, neither a swaggering Macho Man nor a simpering SNAG (Sensitive New Age Guy). What I am is a man who loves women precisely because they *are* women. As one who celebrates diversity—especially the intellectual electricity and (yes!) carnal delights arising from sexual dimorphism, without which life would be a boring institutional gray—I'm no fan of anything that would blunt the delicious natural distinctions between woman and man. To me, elective androgyny is a woeful waste of the priceless passions of complementary opposites: the cosmic bliss of yin and yang.

In league with Paul Shepard, I base my biological views of gender roles, in hunting and elsewhere, primarily on personal observation and logical deduction. Considering the excitingly empirical, warmly palpable physical distinctions between the sexes, along with more subtle psychological/emotional differences, it's hard to deny that women are biologically better designed and psychologically better suited than men for certain social roles and physical functions—including recreational endeavors. Men, likewise, naturally gravitate to male-specific roles, functions, and recreations.

Traditionally—with and without cultural conditioning—hunting has always been overwhelmingly a male passion . . . which strikes me as entirely natural, and thus to be expected, if not always accepted. Yet male "domination" of hunting should never be enforced, directly or obliquely, which too often it has been and still is. Happily, more and more women of late are saying to heck with that and daring to "cross the line" and take up hunting. And I say: More power to them.

While I recognize and praise gender *distinctions,* I condemn social stratification or personal domination based on sex. Paul Shepard notes: "Male chauvinism based on evolutionary determinism is reprehensible. But it is futile to pronounce, at the end of fifteen million years of hominid evolution, that men and women are alike."

Not only futile, but dismally unimaginative. Where gender distinctions become a problem, a source of conflict, a form of persecution, is through socialized sexism. Because "might" so often passes for "right" in this imperfect world—and men possess the bulk of physical might, magnified by hormone-fueled aggression—men historically, across the relatively recent interlude of recorded history, have forcefully, often brutally, lorded it over women.

In any sane and logical society, sexual dimorphism would define synergy, not tyranny. But where are social sanity and logic today? Most of what remains of both—say Shepard, *Ishmael* author Daniel Quinn, and a growing number of informed others—resides where it originated: with traditional foraging peoples. In *The Tender Carnivore and the Sacred Game,* Shepard sums up the accepted ethnological knowledge that "among living ['primitive'] hunters, men make the major decisions relating to their activity as hunters, explorers, and symbolizers, while women dominate in matters of children, diet, and social relations. An overall gender-based, dominance-subordination ranking does not exist among them. Hunters are not patriarchal authoritarians like cattle- and sheep-keepers. They perceive gender as the outcome of a cosmic philosophy of complementarity."

Mary Zeiss Stange concurs: "It seems clear . . . that for upwards of two million years humans lived in remarkably stable—and probably fairly egalitarian—hunter-gatherer cultures. . . . The first clear archaeological evidence of social or class stratification does not occur until the sixth millennium B.C.E. in Neolithic Mesopotamian farming villages."

If the hunting lifeway indeed has a multimillennial history of gender equality, then why are so many modern male hunters so gender-jealous of "their" hunting heritage? Because, I propose, men who feel, speak, and act gender-possessive of hunting do not embody an authentic human hunting tradition. Rather, they are metaphorical "cattle- and sheep-keepers" who happen to hunt. It's no surprise that the bulk of men who feel that hunting should be exclusively "a guy thing" belong to Stephen Kellert's least-admirable dominionistic/sport subset. Most males who welcome women into the hunting ranks and, as important, view them as equals, by contrast, fit Kellert's utterly admirable naturalistic/nature hunter type.

One of the many important lessons I've learned from my wife Caroline—a reservoir of commonsense wisdom—is that not only jealousy but arrogance and aggression arise from ego insecurity. Like frightened dogs, insecure people, mostly male but often female, are doomed to bull through life trying to bluff others, and themselves, into believing they are strong and worthy of admiration. Macho male hunters fit this shoe snugly.

The work of Stephen Kellert, Mark Duda, and others has substantiated that for many men who hunt—especially those of the dominionistic/sport persuasion—"proving" their manhood by "testing themselves" against wild animals and wild places holds most of the attraction. So when a woman takes on that "manly" challenge—especially if she succeeds—she is slapping the wind right out of that Hemingway myth of Man the Heroic Hunter . . . prompting the macho ego to sag like a wet flag.

In desperation, gender-threatened male hunters often retreat to the brainless bunker of "tradition." As in: "Hunting *traditionally* is a male activity and should stay that way"—reminding us that humans evolved as "man the hunter/warrior; woman the gatherer/nurturer."

On both counts—tradition and social evolution—they are wrong. Let's begin by asking: How many generations of doing a thing does it take to establish a "tradition" in a national mind that's little more than two centuries old? The sacrosanct western ranching "tradition," for example, rarely runs deeper than a handful of generations, a century or so, during which brief time it helped to destroy a Native American hunter/gatherer tradition *hundreds* of generations and *thousands* of years old.

Exactly likewise, the "tradition" of hunting as an exclusively male endeavor is largely a myth based, not on deep-time history, but on postagricultural paternalistic convention. Shepard observes: "Hunting in primal foraging societies has never excluded women. Their lives are as absorbed in the encounter with animals, alive and dead, as are the men's. . . . Traditionally the large, dangerous mammals are usually hunted by men, but it has never been claimed that women only pluck and men only kill. . . . Paleolithic female figures occur in sanctuaries where the walls are painted with hunted game."

Would this be so—would women be represented in these graphically sacred Paleolithic shrines—if they had been excluded from the "big game" hunting tradition these shrines honor? In fact, hunting became a self-avowed male possession only with the agricultural "revolution," concurrent with the advent of other sexist aspects of patriarchy that many men (and a few women) cling to as normal today, but which were (so far as can be determined) aberrant throughout the Paleolithic millennia.

Even the stodgy world of anthropology—enamored of its own "traditions" and traditionally male dominated—is coming around and admitting the flaws of the rigid "man the hunter" paradigm—as evidenced recently by a *Discover* magazine feature whose title and abstract proclaim:

> NEW WOMEN OF THE ICE AGE: Forget about hapless mates being dragged around by macho mammoth killers. The women of Ice Age Europe, it appears, were not mere cavewives but priestly leaders, clever inventors, and mighty hunters.

In this eight-page cover story, author Heather Pringle calls upon the research, discoveries, and compelling theories of a trio of respected American archaeologists to refute the idea of Pleistocene life as "the ultimate man's world."

These researchers—led by Olga Soffer—propose that life in Ice Age Europe (and, by implication, Ice Age Everywhere) "had little to do with manly men hurling spears at big-game animals." Instead, says Soffer, Pleistocene survival, according to evidence unearthed at several sites, "depended largely on women, plants, and a technique of hunting previously invisible in the archaeological evidence—net hunting."

"Net hunting," says Soffer, "is communal, and it involves the labor of children and women. And this has lots of implications." And those implications all lead to a logical, if still yet theoretical, deconstruction of long-held ethnographic traditions regarding Pleistocene gender roles. In the past, Pringle says, "most researchers ruled out the possibility of women hunters for biological reasons. Adult females, they reasoned, had to devote themselves to breast-feeding and tending infants."

This sounds entirely reasonable. But is it?

Ethnography is the study of contemporary cultures, often in search of insights into similar cultures of the past. By studying living hunter/gatherers, ethnographers are able to draw certain loose parallels with vanished cultures who subsisted by similar means in similar environments. "But when researchers began turning to ethnographic descriptions of hunting societies," writes Pringle, "they unknowingly relied on a very incomplete literature. Assuming that women in surviving hunting societies were homebodies who simply tended hearths and suckled children, most early male anthropologists spent their time with male informants. . . . 'When they talked about primitive man [here Pringle quotes Soffer], it was always *he*. The *she* was missing.'"

In fact, Pringle notes, recent research reveals that "women and children have set snares, laid spring traps, sighted game, and participated in animal drives and surrounds—forms of hunting that endangered neither young mothers nor their offspring. . . . They even hunted, on occasion, with the projectile points traditionally deemed men's weapons."

And in their net hunting, these "cave" women captured big game as well as small. Soffer notes: "Accomplished weavers in North America once knotted mesh with which they captured 1,000-pound elk and 300-pound bighorn sheep." And likewise, it now appears, in Europe.

A few winters ago, I participated in the netting and ear-tagging of Colorado mule deer for a habitat utilization study. With nearly two dozen does, fawns, and antlered bucks feeding beneath a rope net suspended on poles above an alfalfa bait, the net was dropped by remote control. We human "hunters," male and female, then rushed in and were able to lay hands on the animals, one on one, and gently wrestle them to the ground. Once down, they rarely struggled further, so that we were able to examine, tag, and release them without harm to them or to ourselves. From this experience, I can verify Soffer's assertion that once even a large animal is tangled in a net, virtually anyone, whatever the age, gender, or strength, could safely kill it—using clubs, spears, or even rocks.

In addition to nets, subsistence hunters also constructed stone, wood, and brush fences to channel game toward pitfalls—as described by Siberian explorer V. K. Arseniev in *Dersu the Trapper*. Here again, the coup de grâce could be safely delivered by anyone.

Another form of Paleolithic big-game hunting, where geologically practical, was the "buffalo jump"—particularly popular among prehorse North American Plains tribes. Typically, all able-bodied members of the clan or tribe would participate in stampeding and channeling a herd of bison or other animals over a cliff to their death.

But whether Paleolithic women actively hunted large mammals or not, a central element of the evolving "woman the hunter" counterparadigm is that big critters rarely comprised the bulk of any primitive people's diet anyhow. Toward this end, Heather Pringle relies on the work of lithic (stone tool) expert Linda Owen to make her point that a majority of the human diet during the European Ice Age (and, by inference, the North American Pleistocene as well) consisted *not* of such massive megafauna as mammoth—as the "man the hunter" tradition maintains—but of smaller animals, insects, and plants.

Certainly this is true for virtually all surviving hunter/gatherer cultures today, even though all such peoples have been cornered in harsh, often human-devastated, ecosystems where few large prey survive. If a similar diet paradigm was universal for preagricultural foragers, says Pringle, making a risky leap of logic, then "it was women, not men, who brought home most of the calories to Upper Paleolithic families. Indeed, [Owen] estimates that if Ice Age females collected plants, bird eggs, shellfish, and edible insects, and if they hunted or trapped small game and participated in the hunting of large game . . . they most likely contributed 70 percent of the consumed calories."

Owen's optimistic calorie counting—and Pringle's hopeful endorsement of same—smells more like hopeful speculation than science to me. But whether Owen's caloric contentions are correct or not, Pringle's core argument—that back when all human cultures were foraging cultures, women as well as men were active hunters as well as gatherers—is both scientifically and logically likely. Consequently, for those in search of a true and prime "human hunting tradition," it's not "man the hunter," not Artemisian huntress cults . . . but egalitarian cooperation.

———

MONTANA LITERARY LEGEND A. B. Guthrie, Jr. liked to say that if he could rewrite Genesis, it would open: "In the beginning, there was the Word. And the Word was *change.*"

While perhaps not of cosmic import, the most exciting change taking place in hunting today is the rapid growth and gradual male acceptance of female participation. "The relative stability in the percentage of Americans who hunt masks a very important interaction by gender," professes Responsive Management. "Hunting among American males decreased between 1980 and 1990 while hunting participation among American females . . . more than doubled between 1985 and 1990, from 1.3% to 2.7%." According to *Time* (November 30, 1998), this trend continued through the 1990s, when roughly twice as many American women, some 2.6 million, annually purchased hunting licenses as had in the 1980s. Why?

"The increase in participation in hunting among females," Responsive Management (somewhat inconclusively) concludes, "appears to be due to social changes." Happily, these amorphous "social changes" include not only the increasing acceptance by men of women as hunters, but a true and sincere male enlightenment as well. More and more men, myself included, are learning that women can be *good* hunters and *good* companions afield. And many of us are coming to truly enjoy, even prefer, their company.

Even so, some male hunters, having been socialized to accept paternalism, tend to view women hunters as perpetual trainees. When young male nimrods grow up, they are perceived as veterans whether they know beans about hunting or not; meanwhile, women hunters too often are viewed as a priori immature in woodcraft, no matter their obvious skills. Part of this is simply wishful thinking: some men welcome women as hunters only so long as they remain unskilled relative to them (and thus no ego threat). To illustrate this common male dis-ease, allow me to tell an anecdote on good ole Erica.

A "gnarly ranch girl" long before she became a "gnarly mountain chick," Erica is a master at field-butchering large animals. While elk hunting a few years back, she came upon a couple of gents who had killed and field-dressed a cow elk but hadn't a clue how to quarter it for packing out. Erica, who could do such work in a rainstorm at midnight, generously and gently (being sensitive to the tender male ego) offered the flustered fellows some suggestions on how to proceed.

The boys listened politely, without argument or protest, but clearly suspected any such "manly" advice proffered by a "little lady," opting politely to await the arrival of their guide. Fine, said Erica. But before she could get away, the guide rode up and set to work on the elk, carefully explaining, step by step, how it was done—precisely as Erica had just outlined. Visibly embarrassed, the greenhorns rolled their eyes and shuffled their shiny new cowboy boots, but said nothing. Stifling a smile, Ms. Fresquez turned and walked away, continuing her solo hunt.

For myself, having dabbled in guiding (for the experience more than the money), I've come to believe that men who feel at ease hunting with woman guides are rare birds indeed. This is less a matter of male lack of confidence in female competence and more a matter of socialized gender pride. After all, whether he realizes it or not, any man who has to hire professional help in order to hunt successfully—the oldest and most basic of male skills and social expectations—must to some degree feel his ego threatened. Consequently, to accept the leadership of a woman hunter, especially in view of other males, would be for some men just too much.

Yet, I believe that little of the widely perceived male bias against women hunters is truly sexist—that is, mean-spirited, domineering, and intended to "keep women in their place." To the contrary, I sense that most men (including, maybe, Erica's father) who uphold the sanctity of the all-male hunting camp—that most hallowed of traditions among men who hunt—are merely seeking an opportunity to get off by themselves for a while in order to indulge in saying and doing "man" things in the comfortable company of other men. This strikes me as neither patriarchal, sexist, nor insecure, but natural and even healthy. As feminist scholar Florence Shepard recently remarked in conversation: "Both men and women like to do same-sex things together, enjoy each other's company, and have their own kinds of gatherings. I think there is a need for all-male and all-women get-togethers. Although I like men, I love to be with a group of women, and my interaction with women alone is different than in a mixed group: I feel more relaxed and can talk of womanly things that would seem silly or trite to men. The same is true for men. Same-sex activities are both legitimate and important."

———

WHICH REMINDS ME of a story I heard a few years back, across a timberline campfire, from a Colorado game warden.

Glen Eyre was riding backcountry patrol during elk season when he noticed a fresh line of horse tracks leading off into the woods where no trail existed. Curious, he followed the spoor to a secluded camp. Approaching quietly, Glen looked the scene over. Eight horses were tethered nearby, grain bags on their noses. No game was hanging or otherwise evident, and it appeared no one was home. Then he heard the laughter. From the sound of it, several men were playing cards inside a big wall-tent.

Announcing himself as he rode into camp, the warden was greeted by four men—two older, two younger—emerging from the tent. It was mid-morning, but clearly they'd been drinking, though "not to excess. And from the raggedy looks of them," Glen told me, "they'd been at it all night."

The hunters invited Glen in for coffee, which offer he accepted, even as he asked to see their hunting licenses. All seemed in order. Half an hour later, as the warden was leaving and the men resumed their poker, Glen asked two nagging questions: "Why aren't any of you fellows out hunting this fine morning? And where are your rifles?"

"Guns?" replied the oldest man, chuckling. "We left them in the truck. Don't want no *guns* around camp when we're *drinking!*"

And then the whole comical story came spilling out: This quartet—two old brothers and their adult sons, all from Arkansas—had been playing this charade for years. Though they usually hunted some, none of them really came for the elk. "The terrain's too tough on us lowlanders," they explained. Even so, they loved the big blue mountains and the cold clear mornings and the wilderness and camping and, especially, the time alone together.

And—here's the heart of it—they feared that if they admitted to their wives how little time they spent actually hunting, they'd want to come along and all the "guy fun" would end.

In reality, I'll bet their wives, whether "the boys" wanted to admit it or not, were just as happy to have some time to *them*selves.

—————

ONE FINAL ANECDOTE: During the elk season just past, Erica and I hired on with outfitter T. Mike Murphy for a week in his wilderness hunting camp— I as a guide and camp flunky, Erica as cook and wrangler. Six men and gorgeous, young Erica. No tension. No dissension. Every man there agreed— her exquisite cooking aside—that Erica's cheerful and energetic female presence greatly enhanced the experience. This particular group of fellows, it seems, had no real need for male bonding and guy talk. We had come to hunt, and Erica, as an experienced hunter, fit right in. Every man there was happily married (just as Erica is happily unmarried), so there was no flirting, no gender jokes or pranks, no disruptive sexual tension.

This is not to say there should never be sexual tension in a hunting camp. Like other experiences, hunting is sometimes better alone. Sometimes it's better in company. And sometimes, it's best of all when that company is someone you can warm a sleeping bag with. In praise of this latter scenario, the subtitle of a recent *Bugle* article proclaims: "Hunting camp can be exotic and romantic and strengthen the bonds of love. Take her there." Yes, please do—though I would add the cautions: Only if she *wants* to go, and just the two of you, *alone.*

Ironically, some of the same women who resent their men going off to hunting camp have no personal interest in hunting or even camping—just as some men who object to activities that take their wives away from home would be disinclined to tag along, even if invited. But these are psychological, not sociological, phenomena and in no way universal.

Caroline, for a central instance, loves camping and all manner of outdoor adventure but cares nothing for big game hunting. (Chasing grouse is a different story.) Consequently, I hunt elk, deer, and pronghorn mostly alone, occasionally with other men, often with our mutual good friend Erica—and yes, I too have come recently to enjoy an annual few days of "guy fun" in an all-male hunting camp. And in all of this, I receive Caroline's blessings.

And so it goes, round and round, forever.

LIKE IT OR NOT, more women are hunting today. And most of them, whether by choice or necessity, hunt most often in company with men. And so far as it goes, this is good. Yet I'd like to see more Ericas out there as well— women who choose, often as not, to go it alone or in company with other women, thus immutably confirming their "taking back." Gender be damned.

Working hard to promote just such a happy eventuality is Wisconsin professor Christine Thomas. According to Mary Zeiss Stange, Ms. Thomas perceived a decade ago that "women's nonparticipation in traditionally male-identified outdoor activities often had less to do with lack of interest than with negative peer pressure, lack of access and encouragement."

To change all of that, in 1991 Thomas and friends organized the first Becoming an Outdoors-Woman (BOW) workshop. It attracted more than a hundred women of impressive diversity: aged sixteen to sixties, rich and poor, poorly and highly educated, single, married, divorced, with and without children. "What united them," says Stange, "was their desire to learn hunting and fishing skills in the noncompetitive, supportive atmosphere the workshop planners had promised." This meant mostly, if not exclusively, female instructors.

And that was the start of something big. Across the next five years, BOW graduated twenty thousand women from workshops in three Canadian provinces and forty states. And keeping BOW going and growing, many of the graduates—after honing their skills in hunting, fishing, shooting, archery, boating, camping, land navigation, wilderness survival, and more—reciprocate by becoming BOW instructors themselves.

My personal experience with BOW is limited to this: In the course of an Ontario Hunting Heritage/Hunting Futures conference and workshop I attended in 1998, noted Canadian outdoorswoman Freya Long was honored

by the Ontario Federation of Anglers and Hunters for her labors in organiz-
ing and overseeing a local BOW chapter. In her acceptance speech, Ms. Long
recalled a conversation with a recent BOW graduate who called to say how
excited she was about the upcoming hunting season, the first in which she
would participate:

> She told me that she and her partner talk frequently about
> what hunting means to him, and he explains to her about
> the emotional side, the way he *feels* about hunting. He
> explains to her about the adrenaline rush he gets when the
> days get shorter, the air grows crisp and it's time to go out
> scouting. And just this week . . . a flock of geese went over.
> She stopped. She looked to the sky. She heard the honking.
> And she felt something special, something she had never
> experienced before. She said to me: "I got the adrenaline
> rush he told me about, Freya. I felt the thrill for the first
> time; I am starting to know what he means about hunting."

Today, when I see women afield, as I increasingly do, I'm filled with the same
hopeful optimism for hunting's (and humanity's) future I've always felt when
I see a father and son or grandpa and grandkid hunting or fishing together,
each lost in love and appreciation for the other and their shared experience
in nature.

In the years ahead, I hope to see more mothers and grandmothers out
hunting with their kids as well—not just "taking back," but *passing along* a
true tradition.

16

Should Kids Hunt?
Reflections on the Past
and Future of Hunting

> Does it help them connect with their elders and the out-
> doors; to respect the power of weapons and the realities of
> life and death, as hunters believe? Or does killing animals,
> as hunting's opponents claim, damage young psyches, mak-
> ing children indifferent to suffering and ready to see deadly
> violence as acceptable behavior?
>
> —Lance Morrow

IT'S TRUE WHAT THEY SAY—that recruitment of young people to hunting
has fallen off in recent years. Some cheer this news, proclaiming that children should not hunt. In fact, one of the favored tactics of "animal rights"
hunt-disrupters is interfering with special introductory hunts for kids. (Like
in-your-face heckling at abortion clinics, this practice is now illegal in most
states.)

Others meanwhile—not only hunters but bio-wise nonhunters as
well—fear that a continued decline in young hunters bodes ill not only for
the future of hunting but for the wildlife and wildlands that hunting helps
perpetuate, as well as for all those young Americans who will never know

the natural joys and personal epiphanies that only true hard hunting can bring.

Why is this happening? And what, if anything, can be done to reverse the trend? Or should it be reversed?

If we want to influence the future, and do so wisely, we must start by looking back.

———

WHEN I WAS A YOUNGSTER just learning to hunt, my challenges were to decipher the daily lives of wild animals and the subtle secrets of the places they live; to master woodcraft and ancient hunting skills; to shoot a rifle steady and true with iron sights only (and a stickbow with no sights at all); to read and rely on map and compass; to interpret the prescience of clouds and the stories writ in animal spoor—and so much more. Until recently, this is how it was for all young hunters, and it all was good.

Perhaps hunters of my generation were motivated as much by a limited availability of gadgets as we were by preference. From the distant vantage of middle age, I can't really say . . . I just know I'm grateful that it was as it was. An old dog now, I still prefer "primitive" hunting tools and self-reliance to store-boughten crutches. And I know of no one who takes more pleasure from, or finds more meaning in, the hunt than low-tech "traditionalists" like me . . . nor, for what it's worth, whose freezer is more consistently filled with wild meat.

Most important, then as now, I am acting on, acting out, and *satisfying* an instinctive yearning for a true, traditional hunting challenge—an ancient, innate itch that can be scratched only by making the least of stuff and the most of self. I believe culturally unadulterated youngsters, whether they know it or not, share this same visceral need to "go wild." And it's an itch that contemporary hunting, with its market-driven focus on off-the-shelf shortcuts and body-count "success," largely fails to satisfy. Techno-hunting is just too much like the rest of modern culture—escaping which, after all, is a prime attraction of all wilderness endeavor.

———

THIS IS NOT AN AUTOBIOGRAPHY. If it were, I'd have bored you from the start with my childhood experiences. I can justify the anecdotes from those for-mative years, which I'm about to drop on you now, only insofar as they pro-vide a personal backdrop for the opinions I'm about to express—and, to a lesser degree, because here, near the end of my thoughts on hunting, they help bring us full-circle, back where we began, providing something approaching closure.

———

BORN AT THE WEEDY EDGE of a large city, from childhood I longed for field and stream, river and lake, mountain and canyon, forest and flat. My father, who worked too much and had yet to learn to relax and enjoy life, was little help. Happily, there were other hunters in the family who were able and, to varying degrees, willing to play part-time mentor.

My maternal uncle Charlie Harper was a reluctant one of these. On cherished rare occasions, Uncle Charlie would allow me to tag along, safely unarmed, as he prowled the fringes of farm field and pasture, his wiry pointer, Dick, ranging ahead, nose working, tongue lolling, hot on the trail of bobwhite quail.

Back home, if I nagged him long enough, Uncle Charlie would bring out rifle or revolver or shotgun, check to assure it was empty, and allow me to handle the steely-blue icon: feel its form, heft its weight, inhale the aromas of cordite and powder solvent, aromas that to this day induce nostalgia. And it was Charlie who first introduced me to the culinary pleasures of wild meat: quail, pheasant, venison.

But on reflection, this reluctant mentor's greatest gifts to me were his retellings of wilderness mule deer hunts in Colorado and elsewhere in the Rocky Mountain West. This quality of outdoor adventure, I knew even then, though I was not yet ten, was something I must experience.

Where Uncle Charlie left off, Cousin Harlis began. If I was ten at the time, Harlis Harper was not yet twenty. On frequent weekends—equipping me, incrementally, as I learned to respect firearms, with a BB gun, a pellet rifle, a single-shot 20-gauge shotgun, and a bolt-action .22 rifle—Harlis would lead out across grassy fields and in amongst thorn-tangled woods in search of wild meat and adventure . . . and some subtle something more.

Harlis was a patient hunter, and he patiently taught me to be the same: walking slowly and quietly, pausing often—standing still as statues, listening, looking, breathing deep the clean country air, then tiptoeing on. When we did spot game, unless immediate response was called for, Harlis (and I by example) would freeze for long memorable moments and just watch: studying, admiring, looking to learn, learning to love.

There were others as well, friends and fathers of friends, who occasionally pitched in. And thus did a precious few mentors, singly and together, expand and bless my life forever. As soon as I was old enough to drive, having worked and saved to buy my first car, I took to going fishing and hunting on my own, at every opportunity—fishing usually with friends, and hunting likewise if birds or small game were the object. But for the serious, magical business of big game hunting, then as now, I went by preference almost always alone.

How clearly I recall the excitement of fall Fridays during archery deer season, when I'd drive too fast for an hour after school to reach my happy hunting grounds, where I'd pursue ghostly, gray-bodied whitetails with an Osage orange longbow and a back-quiver bristling with hand-fletched wood arrows. Come dark, I'd heat and eat a can of stew, then crash in the cramped back seat of my '53 Chevy with coyotes yodeling all around. Then up again before sunrise to hunt all day Saturday. Sunday the same. And I loved it all, as I do yet today, mild misery and prolonged frustration notwithstanding.

It took four years of hard hunting to win my first deer: a plump November white-tailed doe that walked beneath the live oak limb upon which I was tenuously perched, shivering in my boots. My arrow took her through both lungs and she died on the run, just five heartbeats later. I remember being at once elated and saddened, whooping and weeping—a bittersweet emotional tug-o-war I've since come to expect and respect, and which all true hunters know all too well.

I am no longer young. Yet the years fall away and precious memories revive each time I see veteran and initiate afield together—moving slowly, stopping often, talking in whispers if at all, the eager youngster emulating the elder's every move. Harlis Harper and me. Your mentors and you.

———

MENTORS. Even if we've never encountered the term before, we've all had them, bad as well as good. The original mentor was a teacher in ancient Greece, friend to Odysseus, charged with the education of the hero's son, Telemachus, while the boss was off enjoying his odysseys. Not coincidentally, this prototypal mentor's name was . . . Mentor.

A mentor, simply put, is a teacher, a role model, a guide—for better or worse. My own outdoor mentors were both. Consider Mr. Thorp. I was about fourteen when I met Thorp through his son, another Dave, who invited me to join him and "the old man" for a three-day weekend of fishing. I recall the elder Thorp as thirty-something, ruddy of complexion, small of body, intense, indolent, and creatively profane. What Mr. Thorp was good at was catching catfish. What he was bad at, I was to learn, was life.

For openers, what should have been a two-hour drive to a nearby stretch of river, less than a hundred miles away, required most of a day, with the two Daves left to broil alongside a stenching five-gallon bucket of fishbait in a black station wagon while our leader took hourlong "rest breaks" at every roadside tavern we passed. And there were plenty.

When we finally reached the river—appropriately named the Muddy Boggy—Thorp's first and last physical act was to rig a hammock in the shade and drag an ice chest up beneath it. From this prone throne, he instructed Dave and me—expertly, I must say—in such essential bank-poling skills as selecting and trimming thumb-thick osiers for bank poles and attaching just the right length of braided line, just the right amount of wheel-weight lead to nullify the river's current and hold the bait under, then tying on huge treble hooks and gobbing each one, just so, with a catfish banquet of chicken guts and dead minnows.

For two days, while Dave and I patrolled the brush-choked banks of that cottonmouth-infested river, servicing a score of bank poles and lugging our considerable catch back to camp for cleaning and cooling, Mr. Thorp lay plastered in his hammock, a Falstaff always at hand.

At night, between pole patrols by Coleman lantern, we squatted 'round a campfire purposely made smoky to hold the mosquitoes at bay while a rallied Thorp, nocturnal by nature, entertained us with slurred tales of heroic drinking binges, bar fights, honky-tonk angels, and "serious meat fishin'"— a quaint provincial practice involving a rowboat, a Ball jar of moonshine, and a supply of blasting caps—or, for really big jobs, a part-stick of dynamite, which toy Thorp referred to fondly as a "fishcracker." Poaching deer was another favored Thorpean pastime.

The implied moral of Mr. Thorp's boozy monologues, it seemed, was that when we boys grew up, we too could enjoy such manly sport as boozy bar-crawling, toilet-hugging hangovers, whoring, fish-bombing, and jacklighting deer. Sadly, this too is mentoring.

Happily for me, that was the last I ever saw of Thorp. Later that same summer, he and a companion in crime blew themselves to Kingdom Come (or wherever) while "meat fishin'." Though no witnesses survived to testify, stories abounded. My mind's eye pictures a rowboat, an empty Ball jar, a lighted "fishcracker" fumbled and dropped into the bilge, a frantic, drunken attempt at retrieval, a mighty *Ka-whump!* . . .

The lesson being: If a young person has even a modicum of native sense and the will and freedom to call his or her own shots—as I did by distancing myself from the Thorps after that initial misadventure—even the most blatantly negative mentoring can redound to the positive. But what if—like young Dave Thorp—the negative exemplar is your father or someone else in such a profound position of influence and authority that you lack the freedom, or guts, to "just say no"?

What appeared to be just such a scenario unfolded on a recent August morning, the opening of Colorado's pronghorn archery season. In full dark, longbowman Milt Beens and I hied to preselected ambush sites overlooking separate pools in an intermittent creek on public land. As the sun appeared, so did a husky, high-horned pronghorn buck, grazing alongside a dusty road a quarter-mile below me. But watching him briefly through binoculars would be the only pleasure I'd get that morning—for within minutes came the first in an all-day parade of road "hunters."

These weren't motorized spot-and-stalkers—an accepted practice among pronghorn chasers everywhere, insofar as the distances involved are usually considerable, neither depriving the hunter of the challenge of a long stalk on foot nor the prey of a chance to escape. Rather, these were motorized dilettantes, seated comfortably in lawn chairs in the open beds of pickup trucks, shopping for easy meat. In a variation on that loser's theme, the group that was about to spoil the day for Milt and me—and for that big prongy buck—were packed like pickles into a van with big-city plates and the sliding side door open.

Here in Colorado, it's illegal to shoot, with firearm or bow, from any motorized vehicle, moving or stationary, or from or across any public road. Yet little matters the law to the lawless, and when our heroes spotted the big buck, they sped up in an attempt to overtake him. From the back seat, a tall man leaned out the open door, compound bow in hand and arrow ready. As the van neared the buck, the animal veered off perpendicular to the road. That's when the van slid to a stop in a cloud of dust (cowboy style).

As the shooter leapt out, he gave a loud whistle, halting the buck's retreat at something around a hundred yards. (Spooked wild animals often can be stopped, if only briefly, with a yell, whistle, or other loud noise that whets their instinctive curiosity.) Feet firmly planted in the middle of the road, this paradigm of sportsmanship tilted his weapon skyward and mortared a hopeful missile toward the ridiculously distant target.

From my catbird seat on the nearby hillside, eight-power binoculars in hand, I could see all of this clearly. What I couldn't see was where the arrow landed. The buck didn't flinch, kick, or otherwise indicate he'd been hit, but exploded into flight. Flashing back across the road close in front of the van, the prongy rocketed for the safety of the tree-trimmed creek corridor, between Milt and me. There he disappeared and might have taken any of several erosion gullies climbing from the creek to the timbered ridge above. I never saw him again, nor did Milt.

Meanwhile, down on the road, the shooter was yelling "I hit him, I hit him!" plenty loud enough for Milt and me to hear. Now more bodies spilled from the van, including a lanky teenage boy whom the shooter (still yelling, bonkers on adrenaline) addressed as "Son."

Without bothering to search for his arrow or a blood trail, the shooter hurried toward the creek, where "his" buck had last been seen. Son trailed behind. From near Milt's blind, the man directed Son to search up one wash while he took another. The van, meanwhile, rolled on down the road, side door still open, a new shooter in position, resuming the "hunt."

For the next hour, father and Son ran amok along the ridge just above us, the man yelling out queries and commands punctuated by piercing whistles (his specialty, it seemed). I'd had more than enough and rose to leave. When I stepped into the open, the man spotted me and hollered down, "Hey, guy, you seen a big ol' antelope with an arrow sticking out?"

"Two," I replied in sign language, waving a matched pair of erect middle fingers in his direction.

Bottom line and the point being: What was Son's response to all of this? Was he a willing participant, already ruined—or, as it appeared, trapped by the will of a Thorpean father? Did he leave that place with worry working in his guts that *something wasn't right* about this "hunting" business—an anti in the making—or was he fooled into believing this is how it's done?

Either way, Son loses.

And so does hunting.

———

As AN ADULT, I've been largely an outdoor loner. Doubly so when bowhunting for elk—my passion of passions. Alone is the way I do it best, and alone is the way I like it best. Through the 1980s and 1990s, the hundreds of days I've spent ghosting after elk through bright September woods, gloriously alone, have been among the most enjoyable, satisfying, and indelible of my life. For me, companionship is an evening pleasure; during the actual hunt, company distracts and dilutes—like three on a date. Consequently, nobody was more surprised than me when I became a mentor.

I first met Jamin Grigg, then seventeen, at a bookshop reading of *A Hunter's Heart*. Straightaway he impressed me with his love and knowledge of nature, his concern for the welfare of wildlife and wilderness, his ability (rare in teenagers almost to the point of nonexistence) to ignore peer pressure and dance to his own drummer, and, all tallied, his sober maturity; another Harlis Harper, this one. Jamin was an experienced small-game and deer hunter, anxious for his first taste of wild turkey, then on to elk. I volunteered to help.

Following a long, warm April afternoon of walking and calling, Jamin and I finally found a turkey-talker. When the tom came in to my horny-hen yelps, he was in fact three toms—a big boss gobbler and two yearling satellite males, or "jakes." With the trio of nervous birds smack in his face and bunched together in an unshootable wad, Jamin sat tight as any pro. Lying prone several yards behind him, I could not offer advice, but only watch and hope that he wouldn't let the excitement goad him into taking a stupid shot.

Soon enough and predictably, the big tom sensed a rat in the woodpile, spat an alarm putt, and streaked for the cover of nearby brush. One jake followed like a shadow, but the other lagged a little behind. That's when Jamin's 12-bore shattered the silence.

"I can't believe it," my young friend stuttered as he stroked and admired his bird. "My dad has friends who've hunted all their lives and never killed a turkey, and I get one my first time out! This will blow Dad's mind. He's really going to be proud of me."

Dad was proud, of course. And so was I.

Within a week of that hunt, Jamin had mastered turkey calling and was teaching a high school friend. The following spring, he "guided" his father, sociology professor Kalin Grigg, to his first turkey. Today, four years later, father and son hunt turkey and elk together every chance they get, and have never been closer. In the mentoring business, one-plus-one equals many. Thoughtful mentoring, as I've come to learn, makes more and better outdoorsmen—and women. As a bonus, by introducing others to the skills and values of ethical hunting, you strengthen those skills and values in yourself. Students are often the wisest teachers.

⸻

Given such inviting, exciting circumstances as I and so many others enjoyed in our midcentury youth, with hunting grounds and mentors everywhere, what red-blooded youngster, boys in particular, could resist the adventurous call of the wild? And, icing on the cake, hunting back then was a cheap date.

But such traditional scenarios—ceremonies of passage—are becoming increasingly rare across increasingly concreted North America: increasingly difficult to arrange. Today's neophyte nimrods are made to feel they must amass and master catalogs full of expensive toys—in the process neglecting traditional skills and self-reliance, stifling personal satisfaction, and minimizing the twin values of effort and ethics. Increasingly in these frenetic and confused times, in hunting as in life, it's less the trip and more the destination, less the game and more the endscore.

As C. L. "Chip" Rawlins, poet and reluctant meat hunter, recently observed from his home in rural Wyoming: "More and more, for most of the people who come here from the cities during deer season, it appears that hunting is just an excuse to buy a lot of junk, hop in the truck and go cruising backroads with buddies."

In fact, the market's control of modern hunting is prominent among the reasons we're failing to recruit tomorrow's hunters at yesterday's rates: To all but the least among today's youth, not yet wholly warped by our virtual culture, materialism lacks instinctive appeal in the out-of-doors. What they want, what they need, is meaningful personal challenge: rites of personal passage.

⸻

Further stifling recruitment, today far more than yesterday, aspiring young hunters must concern themselves with finding a *place* to hunt as wild nature gets bulldozed farther away from every city, every day; then finding the means and money to get there; then most likely having to share their hard-earned hunting grounds with others, not all of whom are nice people.

Additionally, today's youth must qualify for and purchase a growing number of certificates, licenses, and permits, compete for increasingly limited draw tags—and, perhaps most daunting of all, struggle to explain their "senseless killing" to a suspicious nonhunting public, some of whom view young hunters as incipient schoolyard killers. And most telling, these critics often include family and friends, many of whom, in their nature-estranged innocence, view wild meat as unclean and (as we've seen) associate hunting with foggy Freudian fantasies of sublimated male aggression.

Whatever your age, you can hardly have forgotten the tremendous weight of influence wielded by friends when you were young. No teenager wants to be considered uncool. And that hasn't changed: peer pressure remains the single greatest sculptor of adolescent attitudes and behavior. And increasingly that pressure either goes staunchly against hunting or could not care less.

In a 1997 survey, Responsive Management polled American teenagers regarding their interest in hunting. "Fifteen percent said they were very interested, 16% said they were somewhat interested, and 18% said they were a little interested. Over one-half (52%) of Americans aged thirteen to twenty years have no interest at all in hunting."

And even among those young people who would love to give hunting a try, where do you find a mentor? Most Americans are living in cities, and in many families both parents are working and fewer adults are hunting. These days, Uncles Charlie and Cousins Harlis are increasingly hard to come by.

———

ALL THIS AND MORE CONSIDERED, if you were a young person considering hunting today—short on access, money, guidance, tradition, and, most important, inspiration—would the shallow rewards of techno-hunting justify its growing social and material costs? It's easier to continue watching TV, cruising Main Street, and haunting shopping malls with your friends, fitting quietly in.

In sum: The recent radical shift in hunting emphasis—from traditional, inexpensive, close-to-home personal challenge to pricey, distant, ethically questionable, and spiritually bereft big-ticket gadgetmania—has clearly contributed to the recent drop in youthful recruitment, engendering the disdain not only of the youths themselves but of the "significant others" who in previous generations would have been prime recruiters.

In RM's words: "It takes a hunter to make a hunter. Almost all hunters are initiated into hunting before the age of twenty, usually by a father or other father-figure."

Yet the hunting "community" seems surprised, searching everywhere for excuses, except within.

———

YOU CAN'T DEFEND the indefensible, no matter how loud and long you yell, not even to the ingenuous young. As Paul Shepard warns, modern hunting "is an easy target: the commercialization and perversion of the hunt, the game hogs, the drunks, the shooters of cows, the facades of camaraderie, make the war against the hunt both easy and facile."

The writing is on the wall. Yet few hunters read it and fewer yet heed it, preferring instead—egged on by extreme elements among hunters, hunters' groups, industry, and media—to hunker down in bunkers of denial while chanting those old losers' mantras: "Right or wrong, we won't give a bloody inch!" and "If you're not completely with us, you're completely against us!"

What to do? As a beginning, I offer the following Four C's: Confess, Clean Up, Coach, and Conserve.

Confess hunting's flaws. When so many—including youthful would-be hunters—can see so clearly what ails modern hunting, today's bunkers of denial may become tomorrow's graves. Too often, hunters hear the charge that for one hunter to publicly criticize the ethics and behavior of another is "divisive," even "treasonous." Not only is such an openly censorial attitude profoundly un-American, it's ultimately self-defeating. Hunters have got to understand and make it clear to the nonhunting public that all hunting is not the same—by supporting ethical hunting while decrying abhorrent hunter behavior. This distinction is a necessity for gaining and holding public support. To defend the indefensible is to swell the ranks of both slob hunters and antihunters—at the same time turning off and turning away the ethical youngsters and women that hunting needs to attract. To criticize the bad is our duty to the good. Hunters with the ability to recognize—and the courage to denounce—morally offensive hunter behavior are neither dissemblers nor traitors; they are heroes.

Clean up our ranks from within—thus depriving our critics of the joy of doing it for us. I'm speaking here of putting heat not only on unethical hunters but also on those elements within the hunting industry, media, and organizations who abuse their positions of trust and influence in order to distort hunter values for personal gain, dividing our ranks and eroding our honor in the doing. *These* are the traitors among us, peddling their perverted "war against wildlife" mentality. We must let these spoilers know what we think—by writing letters of protest and withholding our purchases, subscriptions, memberships, and votes. At the same time, we must increasingly support those outdoor marketers, media, and organizations who demonstrate high ethics and openness to essential reform.

Coach. Become a mentor. Find ways to help people of all ages get started in true, traditional hunting—ways that fit your own personality, qualifications, and situation. For some, this means volunteering to teach hunter safety and ethics courses or Becoming an Outdoors-Woman classes. For others, like myself, it's taking young people (and adult women, and even middle-aged men; better late than never) into the woods, one-on-one, and teaching them, by example as well as words, not only how it's done and how it's not, but why. If today's best hunters won't do this essential seeding work for tomorrow, who will? If we don't care, who should?

Conserve wildlife habitat. Looking back across the century, it's clear that no other segment of society has done so much to restore and preserve wildlife habitat as hunters. Yet, most of the hard work has always been undertaken by a handful of committed individuals. Meanwhile, the rest of us do little more than buy the required licenses and pay the levied taxes, sometimes even grumbling about that. For hunting to survive as the great American democratic tradition it has always been, this must change. What good are "hunter's rights" if you find yourself one day without a place to hunt or anything to hunt for? If we can set aside sufficient public-access wildlands—as the Rocky Mountain Elk Foundation and other conservation groups, both hunter-sponsored and cosmopolitan, are striving so heroically to do—we'll always have game to hunt and lively, lovely places to hunt in, as will our children and theirs. And should we choose not to hunt, through the good work of good hunters we'll still have wild animals and wild places to enjoy, each as we choose.

Yet the percentage of hunters (and nonhunters) who belong to mainstream conservation groups is dismal—even as more and more hunters send off contributions to a growing stable of Chicken Little "hunters' rights" groups with scarcely hidden anti-environment agendas. For democratic hunting to survive in North America, hunters must learn to put superficial differences aside and become hard-core conservationists. It sickens me to hear hunters speak of environmentalists as the enemy. Environmentalism, conservation, preservation—these do not equal antihunting. Rather, they defend the very wilderness and wildlife values upon which hunting and fishing depend.

Whether hunting prospers or dies in this new millennium—or becomes an exclusive private-preserve sport for the rich and privileged, as in Europe and a growing number of elsewheres (including Texas)—will be decided not nearly so much by the big scary antihunters as it will by hunters: through the examples we set for our children; through our ability to admit our faults and our willingness to change when change is necessary; and through our commitment to big-picture, long-view wildlands and wildlife conservation—even when, on rare occasion, such efforts may step on a few of our toes.

LAST WINTER, Colorado hunter Lane Eskew took his ten-year-old nephew, Gavin, and another youngster on their first deer hunt—in a snowstorm, as it happened. Lane was bowhunting, and the boys tagged politely behind. A few days later, Gavin sent this letter to his own "Uncle Charlie":

Dear Uncle Lane,

We very much enjoyed hunting with you! This experience will last a lifetime. I hope that we could go hunting again sometime soon. It was better—way better than sitting at home watching the bube-tube all day. Watching you scout around and get down to shoot the deer was awesome. Kool as you would say.

Hunting was one thing, but de-gutting it and skinning it was a whole new experience. Holding a heart in my hand was, well, unexplainable. Touching a stomach—a real stomach, and my hands in the blood, and tonight, when I went out to dinner, I was actually thinking about what I was eating, witch was ribs.

I was also wandering for my own benefits if you could tell me when I could get my hunters safety card, or whatever. Also if you could tell me when I could get a small game license and all of the little details.

The main reason that I wanted to tell you is that I had to thank you.

Call me the next time you need a partner to hunt with! I would even be
glad to go ice fishing, or maybe just even fishing. I guess that I am one of
those nature people.

Your nephew,
Gavin Wisdom
P.S. Hope to see you soon!!!

Thoughtful hunters will cheer my friend's mentoring efforts. To others, however, fearful that his malleable morals have been badly bent by evil Uncle
Lane, young Gavin's letter will prove alarming. As Heidi Prescott, national
director of the strident antihunting group Fund for Animals (FFA), observes
in a recent *Time* story titled "Should Kids Hunt?": "Both sides are going
after the same target—the kids."

Indeed they are. And Ms. Prescott's FFA is in there fighting hard for
what it believes, following the old environmentalist's strategy of thinking
globally while acting locally. Witness FFA's "Jackson Hole [Wyoming] Bike
Bribe," chronicled in a well-researched article written by Jackson resident
Ted Kerasote and published in *Bugle,* the journal of the Rocky Mountain Elk
Foundation. "The ad in the October 8, 1997, *Jackson Hole News,*" writes
Kerasote in his opener, "caught everyone's attention: 'Hey Kids! Save An Elk,
Win A Bike!'"

Background: Each year, the National Elk Refuge, literally *in* the bustling
Old West tourist town of Jackson, holds a "control" hunt to reduce the number of elk migrating from Yellowstone and Grand Teton national parks and
adjacent high country into the federal wildlife preserve for the winter—
upwards of eleven thousand animals some years. That's two to four times as
many of the half-ton (or bigger) ruminants as the refuge's twenty-five thousand acres of irrigated pasture can support, necessitating costly (and, some
fear, biologically and behaviorally dangerous) supplemental feeding. While
only one or two hundred elk are actually killed in most years' control hunts,
many others are turned back. While the hunt often isn't pretty, it helps.

In 1997, as usual, many of the locally cherished control-hunt permits,
issued by lottery, went to eager youngsters. And even as migrating elk were
the young hunters' targets, the Fund for Animals had its sights trained on
the kids, offering in their ad to give "a brand-new mountain bike, up to
$1,000 in value," to the first twelve- to seventeen-year-old who relin-
quished her or his elk license and promised not to hunt that season. To

FFA's great dismay, not one of the seventy teenagers holding hunt permits took the bike bribe. As Ted Kerasote recounts the incident: "The inventor of the ad, Andrea Lococo, 44-year-old coordinator of the Fund's Jackson office, a former philosophy professor from the University of Louisville, Kentucky, and a newcomer to the Rockies, expressed surprise that none of the young hunters had been interested in her offer. 'Were they simply carrying out the desires of their parents?' she asked rhetorically, when interviewed by the *News*."

Well, yes and no. Yes insofar as their parents almost certainly wanted the meat. But no insofar as none were likely coerced into joining the hunt. Kerasote notes: "Virtually everyone I know in the valley hunts elk for meat and has a deep emotional, even spiritual investment in getting an elk in the freezer; elk meat is the Eucharist of this place." (Even as it is of *this* place, my home, nine hundred miles to the south.)

Yet, seemingly undaunted, Ms. Lococo, attempting to import eastern urban values to the rural West, reaffirmed her loyalty to what Kerasote aptly describes as her "mission." She told the *News* that she intended to charge full-ahead in her "battle for the hearts and minds of children," continuing her efforts to convince them that "you can't love what you kill."

To thoughtful hunter/conservationists like Ted Kerasote, a major flaw in the animal rights worldview and strategy is their patent unwillingness to invest any of their bulging antihunting war chests in programs that directly benefit wildlife or wildlife habitat. Here, thought Ted, was a chance to confront that stubbornness head-on in a spirit of conciliation.

"Given that the connection between more habitat and less supplemental feeding seems obvious to all," Kerasote writes in *Bugle*, "I made a proposal to Andrea. As a board member of Orion: The Hunter's Institute, a Montana-based organization dedicated to ethical hunting and its role in culture and conservation, I had been asked to suggest that the Fund For Animals donate the $1,000—which would have been used to buy a mountain bike—to a habitat-purchase or restoration project of their choice within the Jackson Hole ecosystem." When Ms. Lococo registered disinterest in Orion's proposal, Kerasote phoned FFA national director Heidi Prescott in New York City. Ms. Prescott's response was unyielding: "We'll try again with the bike."

In a finale so bold that few outdoor magazines other than *Bugle* would likely print it, a calm Kerasote summarizes his frustration:

Perhaps someday people in the animal rights movement might come to view hunting in the same light that many of them see abortion: as a profound matter of individual choice. As eaters of wild meat, some individuals take life directly, through bullets and arrows; as vegans, others take the lives of animals indirectly, through combines, pesticides, and habitat loss. Neither is irreproachable. They are choices, and like choosing to bear a child or not to bear a child, are different ways of living in an imperfect and imperfectible world.

Meanwhile, the battle for the hearts and minds of young hunters grows ever hotter—as evidenced by the fact that such a heavy-hitter as *Time* would make a cover story of it. That story (November 30, 1998), written by Lance Morrow et al., is arguably the most informed, realistic, and balanced examination of hunting ever to see print in a major mainstream magazine. Above all, it resists the usual temptation to take cheap shots at what Paul Shepard identifies as the "easy target" too many modern hunters make of themselves.

In his introduction, Morrow posits the central question in this philosophical war: "Does [hunting] help [children] connect with their elders and the outdoors, to respect the power of weapons and the realities of life and death, as hunters believe? Or does killing animals, as hunting's opponents claim, damage young psyches, making children indifferent to suffering and ready to see deadly violence as acceptable behavior?"

After presenting a plethora of pros, cons, issues, and examples, Morrow concludes by circling back to his opening. Once again he asks the central question:

But is hunting safe for children's minds and emotions? Does it, as Lisa Lange says, speaking for People for the Ethical Treatment of Animals, "teach children that life is not valuable"? . . . Terri Royster teaches a class in juvenile crime and behavior at the FBI Academy in Quantico, Va., and for years has studied correlations between childhood cruelty to animals and later criminal behavior. While many serial killers were found to have tortured animals as children, she says, she knows of no research that links hunting and violence against humans. . . . Similarly, Ronald Stephens, director of the National School Safety Center . . . has studied characteristics of children who kill at school, and says flatly that "the notion that anyone who hunts is violent is nonsense."

In conclusion, says *Time*'s Morrow, "teachers and counselors report that kids who are taught to hunt responsibly are generally among the more mature and better-mannered—and saner—adolescents in the wilds of modern American culture."

Could this be, I wonder, because these young hunters, with help from their adult mentors, are taking full advantage of Paul Shepard's advice to adopt, in our estranged modern existence, all possible "accessible Pleistocene paradigms"—those genetically engraved map-pieces to a good and natural life?

Could it be that a wisely guided initiation to hunting helps to develop young people's ability to distinguish between a virtual world of cultural made and a living world of nature born?

If so—and I believe in my heart this is true—then we need all the young hunters, and all the old mentors, we can get. They, both and all of them, are the past and the future of humanity.

17

Bullseye: Process versus Project

Man cannot re-enter Nature except by temporarily rehabil-
itating that part of himself which is still an animal. And
this, in turn, can be achieved only by placing himself in
relation to another animal. But there is no animal, pure
animal, other than the wild one, and the relationship with
him is the hunt.

—José Ortega y Gasset

RECENTLY, BY WAY OF CELEBRATING A BIRTHDAY (never mind which one), I
attempted to tally the time I've spent hunting in my life. And sloppy as my
math may be, those blocks of hours, days, and weeks, scattered across four
decades and more, easily add up to years—years actually out there doing it:
scouting, stalking, sitting, watching, waiting, tracking, getting lost and (in
more ways than one) finding myself, camping, packing meat, and all the
other ingredients of a true hard hunt.

By comparison, within those cumulative years of hunting, the time consumed by killing amounts to only a handful of hours if that. Appropriately,
the hunting adventures I enjoyed most as they unfolded, and cherish most
today in memory, have little to do with killing.

IN ORDER FOR THERE TO BE A HUNT, killing must be the goal: it's the instinctive spark that ignites the ancient predatory fire within all true hunters. Yet, I propose, that goal needn't be realized often in order to experience wholly fulfilling hunts. As Allen Jones writes in his memoir, *A Quiet Place of Violence:* "If the purpose of hunting is only to kill an animal, then the process is moot; we contain the technological ability to kill all animals. The *project* may be to kill an animal, but the project is useful only to the extent that it allows us to orient ourselves to the *process.*"

And unless the hunters polled nationwide by Responsive Management in 1995 were conspiring to pull our leg, Allen Jones and I are hardly alone in realizing this. Moreover, RM's findings suggest the possibility that Stephen Kellert's "sport" hunters—who, like pure meat hunters, tend to be more project than process oriented—are losing ground to nature hunters. According to RM: "Most studies indicate that killing game is not as important to hunter satisfaction as many other aspects of the hunt, including the beauty of the area and the possibility of seeing game."

So if you're a hunter who feels you've failed when you don't bring home the meat or horns—or if you're a nonhunter or antihunter who perceives "the thrill of the kill" as most hunters' primary motivation—you're wrong. Consider, please, this triptych of my most recently memorable hunts.

AUGUST 15. I'm out chasing pronghorns again. Bowhunting, alone as usual, a country mile from camp (camping alone as well), I'm lying quiet as a corpse atop the narrow crest of a hogback ridge overlooking a broad, sage-flecked basin. With my trusty East German surplus border guard binoculars stuck in my face, I'm spying on a high-horned buck as he performs a rude and randy rutting ritual biologists call the PUD—paw, urinate, defecate—a territorial scent-marking dance unique to pronghorns.

No way could I sneak from here to there without getting caught: no stalking cover, rocky ground every crunchy step of the way. Nor is the buck likely to climb this ridge anytime soon, contented as he looks, out there among the girls. But by watching patiently today, I hope to learn the boundaries of his territory, the trace of his PUD line, the rhythm of his rounds, and conceive an ambush for tomorrow. And too, being process oriented, I just like watching.

All of a sudden and literally out of the blue, three golden eagles appear overhead: two adults and a full-fledged young of the year. And *low*, barely more than a wingspan above me. Slow as old molasses, I ooze from stomach to back, going for the bird's-belly view. And what a show they put on!

For the next several minutes the three thunderbirds strut their aerobatic stuff—folding and diving, spreading and soaring, hang-gliding and barrel-rolling. And all the while, the youngster keens and whistles with anxiety, or joy, most likely both, as his (or her) folks guide her (or him) through advanced flight training in an updraft so stiff none has to stroke a wing. At the best of times, in rare quiet moments, I can hear the wind humming like electricity through their feathers.

I've long enjoyed dreams of flying, and though I was a military pilot, my dreams involve no machines, but only extended arms and silent soaring. This, watching these big beautiful birds from so intimate a vantage, is as close to realizing the ecstasy of true free flight as I'm likely ever to come. And plenty close enough at that.

Yes, so . . . what next?

I'd love to report some parting spiritual contact between me and those three winged ones, some gesture of bonding—like, say, one gold-brown feather floating down, alighting on my heart. But in truth, they simply flew away—to resume, I presume, their eternal hunt for freedom, adventure, and ultimate truth . . . leaving me to my own.

———

September 10. Elk season has been blowing and going for two weeks now, and though I've invested serious miles and hours, I've seen only two spike bulls; taboo under current Colorado law. A change in tactics—"something completely different"—is in order. Apropos of tonight I plan to sit in ambush in some lovely, likely spot, quiet as thought, until dark.

It's 5 P.M. as I near a ledge above a familiar spring pool . . . and hear splashing just below. Thinking elk, all the old loved symptoms kick in: hyperventilation, jackhammering heart, hands trembling, my inner carnivore salivating.

Dropping to all-fours, I creep toward the ledge's edge, hoping to sneak a peek. But before I get there, the splashing stops, succeeded by the limb-cracking, rock-rolling sounds of something big, slouching up the slope, coming my way.

Adrenaline whines through my veins like a chemical freight train as I shakily string an arrow and try to compose myself. Soon enough, sure enough, a big patch of beige appears above a bush, just below the ledge, and I think, *that's a mighty short elk*. A moment later, the midget elk morphs into . . . a bear! A huge bear, with milk chocolate torso, dark legs and face, and a flaxen saddle across its broad shoulders; long wavy hair, fall-fat and gorgeous—the classic grizzly look.

Alas, as grizzlike as it appears, and as much as I wish it were, this is no grizzer bear. What it is, is the biggest blackie I've ever seen, a monstrous old boar . . . but wait!

Belatedly, two tiny cubs of the year—"COYs" in bear-watcher lingo—come bouncing up the hill. Oh my. Not only is this the biggest black bear I've ever seen—and I have seen some big ones—it's a *female*. Incredible. Yet, there she stands.

And here I squat, just a leap and a bound away.

No sweat, I counsel myself, recalling the words of bear biologist Beck. Tom advises that healthy, truly wild black bears (as opposed to those who've lost their instinctive fear of people through food rewards and repeated non-violent encounters), even sows with cubs, are rarely a threat to adult humans, so naturally timid is the species. Besides, I'm downwind, fully camouflaged, motionless, and kneeling. Surely the sow (like most females) will stroll right on by without noticing me. No sweat.

But even as I'm thinking these optimistic thoughts, the bear, big as a battleship, alters course ninety degrees starboard and—coincidentally I believe, still oblivious of my hunkered presence—heads straight for me. The COYs, their fur wet and muddy from their recent swim, follow close at heel. There's no time left for indecision—in two seconds more, she'll be standing on my toes—so I do a quick-draw . . . reaching not for my bow, but for the pepper-spray canister riding in a nylon holster at the side of my pack. For once, I have my bear spray when and where I need it. In a flash it's out, the safety off, aimed bearward. *Now* there's no sweat.

This sudden flurry of movement wins Big Momma's full attention fast. But rather than charging, as a grizzly sow absolutely would do in identical circumstances, the big brown blackie stops like she'd hit a glass wall, spins broadside, and glares at me through tiny dark eyes.

At the same instant, and with neither grunt nor glance from Mom, the cubs rocket up adjoining aspen trees, there to bawl and thrash about in histrionic panic, like giant freaking squirrels. I study the sow's face, eyes, hackle hairs, posture, searching for body-language clues to her intentions—but find her (like most females) utterly inscrutable.

Enough. To terminate this increasingly tense encounter quickly and, I hope, peacefully, I slowly stand and take a step—not directly away, which might trigger the sow's predatory chase instincts, nor toward her, which would appear aggressive, but at a neutral angle—my pepper spray still aimed bearward.

In response, the sow blows a tremendous *Whoof!*, switches ends, and then goes bulldozing back down the slope, roaring and huffing ferociously, albeit in faceless retreat.

Not about to be abandoned, the cubs de-tree somewhat faster than terminal velocity and go bawling after Mom, having learned by her good example an essential survival lesson regarding proper behavior in close encounters with humans: *Run away!*

I've been blessed with many memorable close bear sightings while bowhunting in camouflage, sitting quietly or sneaking around, acutely conscious of my own sound and scent. But this was the most thrilling because it was more than a mere sighting: it was face-to-face *encounter.* For a few indelible moments, a big bear, and close, was forced to stare me in the eyes, acknowledge my existence, and deal with it . . . with me.

Because of the intensely personal nature of that high-energy encounter, I cherish its memory more even than that of the young wapiti bull I killed two hours later—who came in to water, looked at me but failed to see, and never knew what happened.

SEPTEMBER 24. My year's denouement is past; a primary project accomplished. Yet the *process* of the hunt continues. The aspens are coin-gold now, most of them, and the rut is rocking along full-tilt boogie. It's been two weeks now since I saw the three bears and bagged my winter's meat, yet I'm still *out here:* still "hunting." Can't seem to help myself and don't really even try. The fact that I'm no longer carrying a weapon matters nary a whit to me, nor to the wapiti; we're both still playing the Sacred Game. And in that game, prey needs predator just as predator needs prey. It's this perpetual round of pursuit and evasion that defines wildlife: wapiti, wolf, and me.

This particular perfect evening, by way of indulging my love for sitting quietly in quiet lovely places, I'm squatting in the waist-deep root crater of a giant fallen fir, spying on a recently and well-used elk wallow twenty yards below, steep downhill. In Septembers past, I've made meat here. Tonight I'm merely watching, listening, whiffing the breeze with lighthearted hope.

A couple of hours before dark, a quarter mile east or so, the evening concert commences with a double-barreled blast of bugling. A big-sounding bull—who, given the confidence of his voice, is this year's king of this particular mountain—has been holding forth over there the past several nights, though I've yet to win a glimpse of him. And now, for the first time, he has competition. For forty-five minutes the screaming rages without surcease, the two bulls cursing one another in what I imagine to be a face-to-face standoff. Winner take all (the cows); loser be damned (to celibacy).

Were I looking to kill a big bull, and so near the season's end, I'd risk a sneak on those two boisterous dandies, mutually distracted as they are. And if I did, as per usual, some unseen sentry cow likely would catch me at the outside edge of the herd, bark her foghorn alarm . . . and that would be that. But tonight I'm just a voyeur, so I sit tight and enjoy.

A few minutes more and the bugling hits a cacophonous crescendo . . . then abruptly stops. The shadowy forest goes silent. Fifteen minutes of unnerving nothingness come and go. I'm thinking about going too, when I hear what I live and lurk to hear—a sharp snap of dry limb, crushed under heavy hoof. Moments later I spy a big patch of beige flashing through the gloaming and I wonder: another bear? No, this time it's an elk. A big bull—head held high, antlers laid low along neck and spine—headed in to water or wallow, circling downslope to get the evening breeze in his pretty face.

Arriving, the bull—a 5 × 6 beauty; the antlers not massive but high-beamed, long-tined, and spread-eagle wide—wades right in for a drink. So close are we that I can see every bounce of Adam's apple, every bulge of belly, as he swallows. Elk, like horses, suck water in through their teeth, slurping like a kid with a straw in a near-empty glass. From my nearby position, I can see and hear this clearly, as many times before. The bull is standing square-on broadside, drinking without a care, without a clue. Were I looking to kill, he'd be a dead elk running . . . and not running far. But this is his lucky night—and, in an unaccustomed way, mine as well.

After guzzling with gusto at leisurely length, the bull hoists his head, water diamonds dripping from thin black lips, and gazes bemusedly about. Now the big head dips again—up again—and down for one final chug. Clearly, our boy is cameling up for the long night of lusting to come.

Done with his drink at last, the giant deer stands like a statue of himself in his earthen water dish—the same pool in which my three lovely bears bathed just two weeks ago—apparently pondering nothing, Zenlike, when a shrill, braggish bugle, from off in the east, shatters the memorable moment. Rather than answering the challenge forthwith, as I'd expect, my boy merely kicks the water: *Ker-splash!*

When the distant bugler sings again, my bull stomps and splashes again, wearing something like a pout on his elongated face. Biologists call this sort of diversionary steam-venting "displacement behavior." For some reason, he's resisting the urge to bugle, channeling it off in other directions.

Now, suddenly and without warning, the 5 × 6 flops whole-body down in the shallow water and commences vigorously horning the muddy earth on either side of the pool, left and then right, again and again, as if honing his antlers for action.

And at last it all comes clear: the heated bugling contest of an hour ago and its sudden surcease, the silent approach of this fine but not quite prime-time fellow, the heckling resumption of bugling from the champ, the close-mouthed, frustrated acting-out of the vanquished challenger. I do believe my friend here has just had his big beige butt kicked, possibly literally and certainly figuratively, and is sulking.

Time passes. The 5 × 6 has been entertaining me for a good long while now, night is creeping in, and I'm actually starting to wish he'd just go away, allowing me to slip out unseen before full-black dark. As if to oblige, the wapiti wallows briefly—rocking onto his back and kicking all-fours in the air like a dusting horse—then rights himself and stands, shaking like a dog with a loud *whop-whopping* of loose wet hide.

Now the long-legged bull exits the pool and starts walking—not back the way he came, but up the hill . . . *directly toward me.*

Within moments, as if I'd reeled him in, the bull is standing just an arm's length away, towering head-neck-and-shoulders above me. I could, I realize, hardly believing, reach out and touch this looming beast: knee, nose, maybe even antler.

To count coup on a bull elk in the wild! Who would think it possible? In exactly this spot several years ago, it almost happened with an innocent year-ling cow . . . but this is different; far, far less likely.

A fierce internal battle erupts: I'm dying to go for the coup, yet some sober something—fear, I suppose, though I prefer to call it prudence—keeps my hands inert.

When the huge head dips to nip some grass just inches from my knee, I find myself staring into one big, brown, mirrored orb, giving all-new meaning to the term "bullseye."

But I haven't long to ponder any of this. As a front hoof lifts, I know the beast is about to take another step forward—and that step will put him at the very brim of my pit, if not within. He could even be contemplating browsing the leaf-pattern foliage on my camo shirt. It's now or never for the coup-counting, and I dig for the reckless will.

But rather than a lightning reach up and out—*Gotcha!*—I watch help-lessly as my hands wave defensively in front of my face, and hear myself hiss-ing *"Go away!"*

Just as startled, no doubt, as if I had touched him, Bullseye dips, stum-bles, and lurches forward . . . even as I duck and dodge, knowing I'm as good as dead, or at least severely damaged. But, literally on the brink of dis-aster, the discombobulated deer catches himself, pirouettes, and crashes away, back down the slope to the wallow—where he stops, shakes his head as if to clear it, peers around though not at me . . . and, incredibly, resumes feeding.

Obviously, Bullseye hasn't a clue as to what low-life species of critter it was who moved and mumbled at his knees; as improbable as it seems, he never got my scent. I could possibly even cow-call him back—but why? I'm not out here to harass wildlife but as always, even when I'm looking to kill, to come and go in anonymity.

It's several minutes more and mine-shaft black before my "pet bull" finally crunches off into the night, allowing me to slink silently away, grin-ning so hard it hurts.

No killing this night, but the hunting has never been better.

———

AND THAT'S THE POINT of all three stories: Although no killing took place during any of these episodes—which, you'll recall, I cite as the most exciting and memorable moments from my past autumn's hunts—were I not a hunter, I'd have missed them all. To reiterate: In true hunting, process out-ranks project.

In a 1996 survey, Responsive Management asked licensed elk hunters to rate the importance of "the chance to kill an animal" in their determination of a quality elk hunting experience. An impressive (in fact, surprising) sixty percent said "unimportant." Flipping that question over, they then were asked about the value of "seeing wildlife other than game species" while elk hunting. A whopping ninety-two percent answered "important."

This is what Allen Jones makes clear with his "process versus project" analogy—and what Ortega was trying to say, somewhat clumsily, with his broadly misunderstood and much-maligned assertion that one does not hunt in order to kill, but kills in order to have hunted.

If that subtle but essential distinction remains unclear or unconvincing at this late juncture, well, there's nothing more I can say. My bucket is empty.

Except, perhaps, to offer a dreamy postscript.

A Dreamy Postscript:
The Wapiti's Message

A dream may let us deeper into the secret of nature than a hundred concerted experiments.

—Ralph Waldo Emerson

To PROVIDE A CONVENIENT SUMMARY—convenient, that is, for those Abbey called "literary crickets," reviewers too rushed to read the whole book (this is not a cut; I've been there myself)—I proffer the following nocturnal fantasy. But beware: Things may get a little weird, in that . . .

"I had a dream."

It always makes me uneasy—clogs my B.S. filter and puts me on guard against pending deception—to encounter that phrase in "nonfiction" writing, since it so often introduces a lie. It's a common device—a quick slick slide from drab reality into colorful fiction. Which is to say: Too many writers these virtual days custom-fabricate "dreams" to round out otherwise flat stories. But not always. At least one credible writer I know regularly enjoys (and occasionally is tortured by) rich nocturnal phantasms that clarify and guide events in his waking life while artfully animating his autobiographical prose.

By comparison, my dreams—what few I recall come morning—are seldom more than tattered shreds of nonsense. If they carry any meaning or message at all, it's that I need to go take a leak. You'd think, once in a while at least, that I'd be blessed with a dream that's entertaining, whether sensible or not. But no—not until the other night at least . . . when I had that dream. It was a hunter's dream, making it a good fit in these closing pages, and peculiar enough that it might be of interest to you. But first, a necessary bit of scene-setting.

———

MY DEAD FRIEND ABBEY—man of letters, desert wanderer, slickrock philosopher, romantic pragmatist—spent a lifetime seeking a transcendent merger with nature. He couldn't explain, even to himself, exactly what it was he was looking for, out *there,* amidst all that roaring heat and roasting rock. So he didn't write about it much, merely mentioning now and again in his journals a nagging desire "to walk into the desert and never return. To become one with the landscape. To just . . . disappear."

In 1989 Ed did disappear (naturally). Realizing at last his desire to join body and spirit with the desert landscape, he now lies beneath it, in a hidden grave, in his favorite sleeping bag.

But to the point: As Abbey employed hiking, river running, rock climbing, and other forms of "nonconsumptive" wilderness adventure to further his quest for a physical and spiritual union with what Paul Shepard calls "the Other"—that is, nonhuman wild nature, inchoate as well as incarnate—my own best mentor has been the hunt. As you're aware by now (unless you're a lit-cricket and only started here), it was hunting, early on, that introduced me to wild (as distinct from pastoral) nature and kindled in me an ache for wilderness adventure and natural knowledge. In turn, that adventure and knowledge, gradually accruing, engendered love and (it follows naturally) fiercely protective instincts for the objects of that love. Before I knew it, I'd become what, back then, I referred to as an "environmentalist." Now, it's so much more than that.

Across the years, all of this has intertwined and intensified to the point that today wild nature and the hunt are the rudders that steer my life. Richard Nelson spoke not only for himself but for me and millions more when he wrote that authentic hunting, like nothing else, "brings me into the wild, and brings the wild into me."

I fish, hike, camp, explore, photograph, study, meditate, and fight for wild nature. Yet it's hunting—most especially bowhunting for September elk, through the intense commitment and personal participation it requires—that facilitates my most meaningful intercourse with "the Other." When hunting well, I *am* wild nature. Such a tough-love relationship with nature points up an irony not lost on hunting's critics: the apparent hypocrisy of killing the very creatures one claims to love.

In self-defense, I'm prompted to report that my lethal consumption of wild life via hunting is direct and honest. When I kill, it's always up close and personal—bringing me into the wild. Moreover, it's my own hands that are bloodied in transforming animal to meat. And what I kill, I gratefully eat—bringing the wild into me.

So I hunt hard and honestly, and feel no guilt when I kill. Yet I too have long been in search of a definitive answer to that persistently perplexing love/kill question: a personally satisfying explanation for this sanguine passion of mine. What, exactly, is this ineffable something I want, I *need,* that only the hunter's path approaches?

In a note to me, his poet's tongue firmly in cheek, Jim Harrison once said of himself: "I hunt because I'm not very evolved." I interpret that as a self-compliment—since most humans today are so synthetically *over*-evolved . . . which is to say, *unnatural.*

Yet there's an increasingly anticlimactic aftertaste to my "successful" hunts. More and more I'm coming to realize that the hunt itself is the thing. Process over project (or product)—and killing kills the hunt. But how to have one without the other?

This, then, is the deeply complex psychological backdrop that renders my dream personally memorable and, just maybe, more broadly significant. The weird part rides on the fact that two years after having this dream— which I described in detail at the time to Caroline, to *Bugle* editor Dan Crockett, and to sufficient others to document its place in time—it tried, via the bullseye incident described above, to come true.

———

Alone as usual, stickbow in hand, I'm sitting silently in the familiar white-barked inner sanctum of an autumn-gold grove of quaking aspens, waiting and hoping for elk to come to water.

As usual when I haunt this well-loved place, a stiff hike up the hill from my humble cabin home, I'm hunkered in the knee-deep root crater of a big blowdown spruce, midway up a steep wooded slope. Twenty yards below me, the slope levels into a narrow grassy bench where a seep spring feeds a tiny, short-run brook. I'm fully camouflaged, of course, sitting stump-still, trying to be invisible. The breeze, blowing lightly from spring to me, is steady and good for my predatory purposes.

The afternoon passes peacefully into evening, and nightfall is slipping in. I'm about to call it quits when, just before dark, a beast the color of shadow materializes soundlessly at the forest's edge, across from the spring, forty yards away. After searching the scene for danger and sensing none, the wapiti steps into the open and I see that it's a bull, his antlers like an old-growth oak—the perfection of elkness and the flesh-and-blood (albeit dreamtime) passion of my hunter's heart. The adrenaline surges and I instinctively tense. Fighting back, I will myself to relax: breathing deep and slow, struggling for self-control. The bull peers my way, but seemingly does not see. Apparently assured, he strides intentionally to the spring pool. My fingers tighten on the bowstring, anticipating the moment when he lowers his head to drink and I can safely draw, aim, and shoot. As always at such times, I feel a strong and confusing empathy for my prey. I wait, throbbing with the thrill of this wildly primitive encounter. But—and here's where the dream gets dreamy—rather than drinking, the bull suddenly turns toward me, an inscrutable awareness glinting in his dark marble eyes: Somehow, *he knows*. Any moment now, he'll bolt for the sheltering woods and once again, as most often is the way, my game will be lost.

Yet, going against all animal logic, the bull doesn't run. Rather, without taking his eyes from mine, he turns and strides along the seep line, along the grassy bench, until he's directly below me. Now he turns again and starts uphill. Mesmerized, I abandon any thought of killing.

Slowly, deliberately, the giant deer climbs, closing the gap until he's standing just one step away and in position, if he so wished, to rip my lungs out with a dip-and-thrust of spearlike antlers: he has the power, and I am helpless to prevent it. The danger is real—yet, gazing into those bulging walnut eyes, I sense no malevolence.

Now the bull takes the final step, towering above me, and peers into my bloody soul, communicating some silent something I understand intuitively, like an arrow to the heart, but could never, ever explain. Without taking my eyes from the elk's, I tentatively extend my right hand toward his muzzle, as if to pet a dog. The bull's nostrils flare, black and wide, testing the tendered hand. After a moment, I inch my fingers up muzzle to forehead to antlers, noting with awe that his brow tines extend well beyond my elbow.

After a moment, the bull pulls away, turns, and looks back at me over his shoulder, clearly beckoning. Abandoning my weapons, I stand and follow the ghostly beast down the hill, past the spring pool, and off into the gloaming—suddenly, magically, it has grown quite dark—one hand riding on his broad muscled shoulders. I am a blind man in the wilderness night; the wapiti is my guide. We, this shadowy creature and I, have magically breached the abyss between Human and Other, predator and prey.

Through the bismuth black we wander, my elk-guide and me, to no apparent purpose. Then, just at dawn, the pristine forest opens—suddenly, joltingly, like the parting of a stage curtain—and we step forward into the edge of a mutilated wasteland of mud and stumps. Throughout this vast clearcut is woven a network of freshly slashed roads, along which appear dozens of concrete building foundations: a subdivision in the making.

No workers are present, but their big yellow machines and their stench and devastation are everywhere. I turn away in horror, back toward the sheltering woods . . . but my elk-guide is gone, vanished.

Feeling lost and terrified, I flee into the trees and run until I can no longer see the destruction or smell the diesel stench; run some more; then collapse in near-death fatigue . . . only to wake some time later, curled, fetuslike, in the bottom of the familiar spruce-root crater above my beloved spring—back where it all began, bow and arrows beside me on the ground, the rutty scent of bull elk heavy in the air.

———

OF COURSE, it's only a nightmarish dream. A hunter's chimera—powered, perhaps, by the same innate drive, seventeen millennia ago and more, that moved Cro-Magnon cave artists at Lascaux, and dozens more subterranean galleries throughout Europe, to overcome what must have been an almost debilitating fear of darkness, depth, and mystery to worm their way deep into earth's cloistered womb. There, with their way and their work lighted only by the eerie dim flicker of reindeer fat burning in small stone cups, these "savages" transformed bare stone walls into artfully stunning, stunningly accurate, palpably spiritual polychrome murals: stirringly numinous portraits of bison, aurochs, horse, red deer, reindeer, ibex, and more, sacred prey species frozen forever in the eternal drama of the hunt. Rarely do human figures appear in Cro-Magnon art, a fact affirming the artists' zoomorphic worldview.

When painting and contemplating the similarly surreal and stirring rock-art images of their own hunters' world, contemporary Australian aborigines speak of entering a nonlinear "dreamtime," where the human/animal/spiritual streams merge and flow as one, round and round forever, forming the river of life.

I've thought a lot about all of this, attempting to relate it to my own nocturnal venture into dreamtime. But in the end, whatever the intellectual path I pursue, or the subconscious desires I may (or may not) harbor to the contrary, I come right smack back to the conviction that we will never, in realtime, bridge the perceptual and cognitive gaps that separate—are *meant* to separate—Human from Other, predator from prey. Through incomprehensible ages, natural selection by tooth and claw—what Shepard calls the "upward-spiraling intellectual dance of predator and prey"—created the elusive wapiti. Identical forces made the human animal a hunter. Without such bloody diversity, there could be no evolution. And without evolution, there would be no shaping, no elk, no me and no you; no animate, sentient life in the universe.

We didn't invent the system; nor, therefore, should we attempt to reinvent it—as extreme animal rights advocates attempt to do via biologically bizarre "perfect world" fantasies that see the end of all predation both human and wild. Such a disappointing failure of logic and of spirit!

We live and we die. We must consume, and one way or another (worms, fire, fish, microbes) we shall be consumed. It's the way, you might say, God planned it. And that, I propose, is spiritual unity of the highest order.

Granted: Direct, violent consumption by hunting and killing is no longer essential to our physical survival. But the chase remains rich and essential nourishment to the civilization-starved human soul. To keep the wild in both of us, and both of us in the wild, I must pursue the wapiti, and the wapiti must evade. Where the forces of natural evolution cease to function, devolution ensues. . . . Just look at our own sickly species.

What haunts me now, as we near the end of this long trail together, is my dream's nightmarish, end-of-nature denouement: yet another pristine forest clearcut; more roads, more houses, more people, more "progress." The scene is made doubly horrific because the nature-devastation it portrays and the natural disaster it portends are real, true, and ubiquitous: On any day across the past several years, I could draw a one-mile circle around my Rocky Mountain cabin, and that circle would contain a dozen or more scenes strikingly similar to, often worse than, the one my dream-elk led me to. Worst of all, this same nightmare is increasingly common at the retreating edge of wildness everywhere, every day.

Against this backdrop, the wapiti's message is clear: It is the immutable moral duty of every person whose being is in any way enriched by wild animate life, via hunting or otherwise, to give something in return for all that is taken. The only real and meaningful way I can truly "become one" with the gloriously untamed creatures I so love and depend upon, in so many ways, is by fighting to protect and preserve the natural places that sustain natural wildness in us all—human and beast, predator and prey; he and she; thee, me, we.

Wildness is an unfettered state of genetic naturalness.

Wilderness is the place where wildness thrives.

Neither exists without the other.

When I hear hunters curse environmentalists as the enemy, I am saddened and angered and embarrassed for them . . . and for myself as "one of them." In parallel, the time is long overdue for animal rights crusaders to face the fact that their words and actions prove that they're not nearly so interested in the welfare of wildlife as they are in imposing their own (decidedly unnatural and thus unworkable) moral values on others. Wild animals don't need human "friends." Truly wild animals don't want human companionship. What they need and want is *wildness*.

While hunters' rights zealots and their opposites within the "animal rights" faction squander tremendous financial resources and personal energies fighting one another, the many-headed hydra of all-consuming "progress" is torturing and slaughtering wildlife and butchering wildlands day and night—and laughing, maniacally, all the way to the bank.

Forget the extremists, whatever their cause; you cannot reason with unreasonable people. And for all their sound and fury, they're insignificant. To kill the insatiable monster of greed before it consumes us all, the concerned but silent majority must learn to overlook rhetorical differences and unite.

Conciliance, that's the ticket. A lasting coalition of objective, intelligent, far-seeing lovers of nature's besieged grandeur, bound in a determined political coalition, working through every possible venue on behalf of natural wildness, natural freedom, and the dignity of all life . . . *that,* I've come to believe, is the meaning of my dream; *this* is the wapiti's message.

What is that haunted look on the face of Wyeth's *The Hunter?*

It is love.

It is freedom.

It is hope.

Selected Bibliography

Abbey, Edward. *Down the River.* New York: Dutton, 1982.

————. "Blood Sport." In Edward Abbey, *One Life at a Time, Please.* New York: Holt, 1988.

Amory, Cleveland. "Now, If I Ruled the World . . ." Interview in *Sierra,* May/June 1992.

Arseniev, V. K. *Dersu the Trapper.* Kingston, N.Y.: McPherson, 1996.

Begley, Sharon, and Andrew Murr. "The First Americans." *Newsweek,* April 26, 1999.

"beringer." ". . . walkin'." *Mountainfreak,* Summer 1999.

Bourjaily, Vance. *The Unnatural Enemy.* New York: Dial Press, 1963.

Bubenik, A. B. "An Immodest Proposal." *Bugle,* Fall 1988.

Buss, Michael. "Meditations of a Hunter." *Outdoor Canada,* November/December 1977.

Capwell, Martha. "Meat Your Maker." *Men's Health Newsletter,* March 1992.

Cartmill, Matt. "The Bambi Syndrome." *Natural History,* June 1993.

Clarke, C. H. D. "Autumn Thoughts of a Hunter." *Journal of Wildlife Management,* October 1958.

————. "Moral and Ethical Aspects of Hunting and Angling for Sport." In *Forest Recreation and Wildlife: Fifth World Forestry Congress Proceedings.* Seattle: 29 August–10 September, 1960.

————. "Quality Hunting." Excerpts from "Moose Hunter Ethics," a speech delivered in September 1977 at the Moose Hunter Seminar, Confederation College, Thunder Bay, Ontario.

Council for Wildlife Conservation and Education. *The Hunter in Conservation.* Newtown, Conn.: 1996.

Crockett, Dan. "Hunting Hope." Foreword to David Petersen, *Elkheart: A Personal Tribute to Wapiti and Their World.* Boulder: Johnson Books, 1998.

Davies, Paul. Quoted in Chet Raymo, *Skeptics and True Believers.* New York: Walker, 1998.

Demarchi, Raymond A., and Anna J. Wolterson. "Results of Special Calf Only

Hunting Seasons in the East Kootenay Region of British Columbia." Proceedings, Elk Vulnerability Symposium, Montana State University, Bozeman: April 10-12, 1991.

Downing, Christine. *The Goddess: Mythological Images of the Feminine.* New York: Crossroad, 1981.

Duda, Mark Damian. "An Attitude About Hunting." *North American Hunter,* September 1997.

———. "Survey Says . . . Archery's Sweeping the Nation." *North American Hunter,* Summer 1997.

———. "Why Do We Hunt?" In *North American Hunter.* 1997.

Duda, Mark Damian, et al. *Wildlife and the American Mind.* Harrisonburg, Va.: Responsive Management, 1998.

Free, Ann Cottrell, ed. *Animals, Nature, and Albert Schweitzer.* Washington, D.C.: Flying Fox Press, 1982.

Fresquez, Erica. "It Isn't Always Easy." *Bugle,* November/December 1998.

Gabriel, Caroline. "Wellness Resources: Wild Meat for Wild People." *Inside/Outside Southwest,* October 1999.

Gaede, Marc. "Hunting? Call it Competition." Letter to the editor in *High Country News,* November 23, 1998.

Geist, Valerius. "Quality Deer and Trophy Management in Europe." Faculty of Environmental Design, University of Calgary, Alberta, Canada, n.d.

Giedion, Siegfried. "Constancy, Change, and Architecture." First Gropius Lecture, Harvard University, 1961. (Quoted in Paul Shepard, "The Tender Carnivore," *Landscape,* Autumn 1964.

Gowdy, John M., ed. *Limited Wants, Unlimited Means: A Reader on Hunter-Gatherer Economics and the Environment.* Washington, D.C.: Island Press, 1998.

Gruchow, Paul. *Boundary Waters.* Minneapolis: Milkweed Editions, 1998.

Harrison, Jim. "The Tugboats of Costa Rica." *Smart,* November/December 1989.

———. "Hunting with a Friend." Foreword to Guy de la Valdene, *For a Handful of Feathers.* New York: Atlantic Monthly Press, 1995.

Haygood, Susan. *State Wildlife Management: The Pervasive Influence of Hunters, Hunting, Culture, and Money.* Washington, D.C.: Humane Society of the United States, 1997.

Jones, Allen Morris. *A Quiet Place of Violence.* Bozeman, Mont.: Spring Creek, 1997.

Kellert, Stephen R. "Attitudes and Characteristics of Hunters and Antihunters and Related Policy Suggestions." *Transactions of the North American Wildlife and Natural Resources Conference.* 1978.

———. *The Value of Life: Biological Diversity and Human Society.* Washington, D.C.: Island Press, 1996.

———. *Kinship to Mastery: Biophilia in Human Evolution and Development.* Washington, D.C.: Island Press, 1997.

Kellert, Stephen R., and Carter P. Smith. "Human Values Toward Large Mammals."

In *Large Mammals of North America,* ed. S. Demarais and P. Krausman. Tucson: University of Arizona Press, 1999.

Kerasote, Ted. "Returning Dignity to the Hunt." *Bugle,* Summer 1995.

———. "The Jackson Hole Bike Bribe." *Bugle,* September/October 1998.

———. "What's Wrong with Fair Chase." Unpublished manuscript, 1999.

Langer, Cassandra L. "Man and Nature: A Basic Relationship." In *The Art of Thomas Aquinas Daly: The Painting Season.* Arcade, N.Y.: Thomas A. Daly Studio, 1998.

Leopold, Aldo. *Game Management.* New York: Scribner's, 1933.

———. "Wildlife in American Culture." In Aldo Leopold, *A Sand County Almanac: With Essays on Conservation from Round River.* New York: Ballantine, 1990.

Marchinton, R. L., et al. "Quality Deer Management—A Paradigm for the Future." *Quality Whitetails,* Fall/Winter 1993.

Marston, Wendy. "Deer Diary." *The Sciences,* November/December 1998.

Medewar, Peter. Quoted in Chet Raymo, *Skeptics and True Believers.* New York: Walker, 1998.

Metzner, Ralph. "The Psychopathology of the Human-Nature Relationship." In *The Company of Others,* ed. Max Oelschlaeger. Skyland, N.C.: Kivaki Press, 1995.

Mitchell, John G. *The Hunt.* New York: Knopf, 1980.

Morrow, Lance, et al. "Should Kids Hunt?" *Time,* November 30, 1998.

National Shooting Sports Foundation. *The Hunter in Conservation.* Newtown, Conn.: NSSF, 1997.

———. *Hunter's Pocket Fact Card.* Newtown, Conn.: NSSF, 1997.

NBEF Newsletter. Loveland, Colo. National Bowhunter Education Foundation.

Nelson, Richard. *The Island Within.* New York: Vintage, 1990.

———. "Finding Common Ground." Introduction to *A Hunter's Heart: Honest Essays on Blood Sport,* ed. David Petersen. New York: Holt, 1996.

———. *Heart and Blood: Living with Deer in America.* New York: Knopf, 1997.

Organ, John F., et al. "Fair Chase and Humane Treatment: Balancing the Ethics of Hunting and Trapping." In *Transactions of the Sixty-third North American Wildlife and Natural Resources Conference.* Washington, D.C.: Wildlife Management Institute, 1998.

Ortega y Gasset, José. *Meditations on Hunting.* Bozeman, Mont.: Wilderness Adventures Press, 1995.

Petersen, David, ed. *A Hunter's Heart: Honest Essays on Blood Sport.* New York: Holt, 1996.

———. *Elkheart: A Personal Tribute to Wapiti and Their World.* Boulder: Johnson Books, 1998.

Pringle, Heather. "New Women of the Ice Age." *Discover,* April 1998.

Quillian, Dan. "Deer Don't Die in Bed." *Traditional Bowhunter.* August/September 1999.

Reiger, George. Quoted in Robin Duxbury, "Stupid Antics Give Me Ammunition to Fire at You," *Denver Post,* January 12, 1991.

Sahlins, Marshall. "The Original Affluent Society." In *Limited Wants, Unlimited Means,* ed. John M. Gowdy. Washington, D.C.: Island Press, 1998.

Schultheis, Rob. *Bone Games: Extreme Games, Shamanism, Zen, and the Search for Transcendence.* New York: Breakaway Books, 1996.

Schweitzer, Albert. *On the Edge of the Primeval Forest.* London: Black, 1922.

———. *The Decay and the Restoration of Civilization.* London: Black, 1923.

———. *Civilization and Ethics.* London: Black, 1923.

———. *The Animal World of Albert Schweitzer: Jungle Insight into Reverence for Life,* ed. Charles R. Joy. Boston: Beacon Press, 1951.

Shaw, William Wesley. "Sociological and Psychological Determinants of Attitudes Toward Hunting." Doctoral dissertation, University of Michigan, 1974.

Shepard, Paul. *The Tender Carnivore and the Sacred Game.* Athens: University of Georgia Press, 1998.

———. "The New Mosaic: A Primal Closure." In *Coming Home to the Pleistocene,* ed. Florence R. Shepard. Washington, D.C.: Island Press, 1998.

———. "Reverence for Life at Lambaréné." In *Encounters with Nature: Essays by Paul Shepard,* ed. Florence R. Shepard. Washington, D.C.: Island Press, 1999.

———. Interview with Derrick Jensen, ed., in *Listening to the Land: Converstions about Nature, Culture, and Eros.* San Francisco: Sierra Club Books, 1995.

Soulé, Michael E., and John Terborgh, eds. *Continental Conservation.* Washington, D.C.: Island Press, 1999.

Stalling, David. "Space Age Technology, Stone Age Pursuit." In *A Hunter's Heart: Honest Essays on Blood Sport,* ed. David Petersen. New York: Holt, 1996.

Stanford, Dennis J., and Jane S. Day, eds. *Ice Age Hunters of the Rockies.* Denver: University Press of Colorado, 1992.

Stange, Mary Zeiss. "In the Snow Queen's Palace." In Mary Zeiss Stange, *Woman the Hunter.* Boston: Beacon Press, 1997.

Stouder, Scott. "Wilderness Technology." *Corvallis Gazette-Times* (Oregon), January 3, 1998.

———. "The Power of Story." *Corvallis Gazette-Times* (Oregon), March 28, 1999.

Sullivan, Mark T. "Brave New Whitetail World." *Sports Afield,* September 1996.

Thompson, Michael J., and Robert E. Henderson. "Elk Habituation as a Credibility Challenge for Wildlife Professionals." *Wildlife Society Bulletin,* Fall 1998.

Vance, Linda. "Ecofeminism and the Politics of Reality." In *Ecofeminism: Women, Animals, Nature,* ed. Greta Gaard. Philadelphia: Temple University Press, 1993.

van Zwoll, Wayne. "Hunting: A Part of the Future or a Piece of History?" *Mule Deer,* Spring 1999.

Ventura, Jesse. Quoted in *Playboy,* November 1999.

Wegner, R. In *Quality Whitetails,* Fall/Winter 1993.

Wescott, Howard B. Translator's preface to José Ortega y Gasset, *Meditations on Hunting*, Bozeman, Mont.: Wilderness Adventures Press, 1995.

Wildlife Legislative Fund of America. "WLFA Update," February 1999.

Williams, Brooke. *Halflives: Reconciling Work and Wildness*. Washington, D.C.: Island Press, 1999.

Williams, Joy. "The Killing Game." *Esquire,* October 1990.

Williams, Ted. *The Insightful Sportsman: Thoughts on Fish, Wildlife, and What Ails the Earth*. Camden, Maine: Silver Quill Press, 1996.

Zimmer, Carl. "Carriers of Extinction." *Discover,* July 1995.

About the Author

DAVID PETERSEN and his wife Caroline live year-round in a hand-built cabin at eight-thousand feet in the San Juan Mountains of Colorado. David is the editor of four volumes, including the journals of Edward Abbey (*Confessions of a Barbarian*) and a literary hunting anthology (*A Hunter's Heart*). He has written six books previous to this, all concerned with nature and wildness—from hiking the wilds of the American West (*The Nearby Faraway*) to exploring the secret lives of bears (*Ghost Grizzlies*) and wapiti (*Elkheart*). Revealing his priorities, David Petersen says of himself: "While I could live without writing, I could never live without the things I write about."

Index